Radical Theologies

Radical Theologies is a call for transformational theologies that break out of traditional locations and approaches. The rhizomic ethos of radical theologies enable the series to engage with an ever-expanding radical expression and critique of theologies that have entered or seek to enter the public sphere, arising from the continued turn to religion and especially radical theology in politics, social sciences, philosophy, theory, cultural, and literary studies. The post-theistic theology both driving and arising from these intersections is the focus of this series.

Series Editors

Mike Grimshaw is associate professor of Sociology at Canterbury University in New Zealand.

Michael Zbaraschuk is lecturer at the University of Washington, Tacoma, and visiting assistant professor at Pacific Lutheran University.

Joshua Ramey is visiting assistant professor at Haverford College.

Religion, Politics, and the Earth: The New Materialism
By Clayton Crockett and Jeffrey W. Robbins

The Apocalyptic Trinity
By Thomas J. J. Altizer

Foucault/Paul: Subjects of Power
By Sophie Fuggle

A Non-Philosophical Theory of Nature: Ecologies of Thought
By Anthony Paul Smith

On Philosophy as a Spiritual Exercise: A Symposium
Edited by Philip Goodchild

The Counter-Narratives of Radical Theology and Popular Music: Songs of Fear and Trembling
Edited by Mike Grimshaw

Theology after the Birth of God: Atheist Conceptions in Cognition and Culture
By F. LeRon Shults

Theopoetics of the Word: A New Beginning of Word and World
By Gabriel Vahanian; Foreword by Noëlle Vahanian

Economics in Spirit and Truth: A Moral Philosophy of Finance
By Nimi Wariboko

Retrieving the Radical Tillich
Edited by Russell Manning

Kierkegaard and the Refusal of Transcendence
By Steven Shakespeare

Religious Experience and New Materialism: Movement Matters
Edited by Joerg Rieger and Edward Waggoner

This Silence Must Now Speak: Letters of Thomas J. J. Altizer, 1995–2015
By Thomas J. J. Altizer and edited by Mike Grimshaw

This Silence Must Now Speak

Letters of Thomas J. J. Altizer, 1995–2015

Thomas J. J. Altizer

Edited by
Mike Grimshaw

palgrave
macmillan

THIS SILENCE MUST NOW SPEAK

Softcover reprint of the hardcover 1st edition 2016 978-1-137-54949-5

First published 2016 by
PALGRAVE MACMILLAN

Palgrave Macmillan in the UK is an imprint of Macmillan Publishers Limited, registered in England, company number 785998, of Houndmills, Basingstoke, Hampshire, RG21 6XS.

Palgrave Macmillan in the US is a division of Nature America, Inc., One New York Plaza, Suite 4500, New York, NY 10004-1562.

Palgrave Macmillan is the global academic imprint of the above companies and has companies and representatives throughout the world.

ISBN 978-1-349-71519-0
E-PDF ISBN: 978–1–137–52249–8
DOI: 10.1057/9781137522498

Distribution in the UK, Europe and the rest of the world is by Palgrave Macmillan®, a division of Macmillan Publishers Limited, registered in England, company number 785998, of Houndmills, Basingstoke, Hampshire RG21 6XS.

Library of Congress Cataloging-in-Publication Data

Altizer, Thomas J. J.
[Correspondence. Selections]
This silence must now speak : letters of Thomas J. J. Altizer, 1995–2015 / by Thomas J. J. Altizer ; edited by Mike Grimshaw.
pages cm.—(Radical theologies)
Includes bibliographical references and index.

1. Altizer, Thomas J. J.—Correspondence. I. Grimshaw, Mike, 1967– editor. II. Title.

BX4827.A47A4 2016
230.092—dc23 2015031697

A catalogue record for the book is available from the British Library.

Contents

Series Preface

Radical Theologies encompasses the intersections of constructive theology, secular theology, death of god theologies, political theologies, continental thought, and contemporary culture.

For too long, radical theology has been wandering in the wilderness, while other forms of theological discourse have been pontificating to increasingly smaller audiences. However, there has been a cross-disciplinary rediscovery and turn to radical theologies as locations from which to engage with the multiplicities of twenty-first century society, wherein the radical voice is also increasingly a theologically engaged voice with the recovery and rediscovery of radical theology as that which speaks the critique of "truth to power."

Radical Theologies reintroduces radical theological discourse into the public eye, debate, and discussion by covering the engagement of radical theology with culture, society, literature, politics, philosophy, and the discipline of religion.

Providing an outlet for those writing and thinking at the intersections of these areas with radical theology, *Radical Theologies* expresses an interdisciplinary engagement and approach. This series, the first dedicated to radical theology, is also dedicated to redefining the very terms of theology as a concept and practice.

Just as Rhizomic thought engages with multiplicities and counters dualistic and prescriptive approaches, this series offers a timely outlet for an expanding field of "breakout" radical theologies that seek to redefine the very terms of theology. This includes work on and about the so-labeled death of god theologies and theologians who emerged in the 1960s and those who follow in their wake. Other radical theologies emerge from what can be termed underground theologies and also a/theological foundations. All share the aim and expression of breaking out of walls previously ideologically invisible.

Introduction

Mike Grimshaw

In 1966, in the midst of what become known as the "death of God" debate, Thomas Altizer stated: "If there is one clear portal to the twentieth century, it is a passage through the death of God, the collapse of any meaning or reality lying beyond the newly discovered radical immanence of modern man, an immanence dissolving even the memory or the shadow of transcendence."[1] Altizer has spent the past half-century thinking, writing, talking, and debating just how we may understand the twentieth century, and now the twenty-first, in the wake of the death of God. As Mark C. Taylor describes him in his foreword to Altizer's memoir: "Thomas J. J. Altizer is the last theologian. As such, he is the most God-obsessed person I have ever known."[2] As the last theologian, Altizer has been determined to compel us to acknowledge that to live in the modern world is to have to continue to think through the death of God. A theologian of the death of God may sound an oxymoron to many, but Altizer is also a self-described *apocalyptic* theologian[3]—an apocalyptic that is simultaneously an ending and an absolute beginning. For Altizer the hope is that we can come to know the dark apocalypse embodied in a nihilistic world "as a joyous apocalypse, and one promising if not embodying an absolute transfiguration" (LDG 177).

What might it mean to think of Altizer as a theologian of transfiguration—a transfiguration of nihilism, a transfiguration of the world when the nihilism of the death of God opens us up to a new possibility, the transfiguration of hope? Central to this is naming the silence—above all naming the silence about God—and this is what gives this collection its title. It is a phrase borrowed from the letter of October 13, 2005, wherein Altizer names this as his new focus: the silence of the abyss of God. This is a theological and intellectual journey; in fact, for Altizer this has entailed a type of existential journey of the whole person into the encounter with what he names "the absolute darkness of the absolute nothingness of God," and this involves a saying yes to God "and hence saying Yes to absolute darkness and absolute nothingness itself" (LDG 179). This is Altizer as theologian of *coincidentia oppositorum* and hence the last theologian, the theologian of the radical transfiguration in the abyss, in the absolute darkness, in the absolute nothing where we say Yes to God. As Altizer proclaims, by "naming the darkness of God we precisely thereby name an ultimate transfiguration" (LDG 180). To be able to participate in this transfiguration, the theological "task is to name that darkness" (LDG 179), a naming as proclamation, a naming that is preached by the self-described "last truly Southern preacher" (LDG 181)—as all of those privileged to hear Altizer speaking publicly can surely agree.

Are these also therefore the letters of a preacher of the death of God? I would argue they often are: they are proclamations reminiscent of a sermon that declaim and name, that challenge and urge us, in a deeply Protestant fashion, to reflect as individuals on what we encounter, an encounter with the abyss that is, again in a truly Protestant fashion, that of the individual with the abyss of God. Altizer notes that a homiletic approach has been central to his theological work (LDG x), an approach manifest not only in his oral communication but, as experienced here, in his written letters.

Altizer is not only the last theologian, but also one of the last true intellectuals, a radical intellectual whose knowledge is the result of wide and deep reading across the disciplines of religion, theology, philosophy, history, and literature. His work reminds us that to do theology, to be a theologian, has required nothing less than an engagement with the history, thought, and expressions of Western culture itself. For that culture is the culture of the question of God, and more recently, the culture of the death of God.

Altizer is radical in two main senses: his work and thought take us back to the *radix*, the root in the Bible, in philosophy, in theology, and it is radical in the secondary sense that it proclaims an alternative, a necessary alternative, to what is taken to be normative. Central to this radical alternative is the work of William Blake and Friedrich Nietzsche, both of whom serve as the *radix* of Altizer's theology, a radical theology that is also, as Altizer often notes, given full expression though the death of God as expressed and experienced in America. So Altizer is a radical theologian, a death of God theologian, the last theologian—and importantly, an American theologian. His theology could have arisen in no other place than America, an America that situates itself always in relation to—and often in competition with—the old world of Europe. An America open to the influence of Asia, an America that looks westward across the Pacific and so remains open to Buddhism in particular. An America that looks to Europe as a place of departure: as an old world and tradition transformed in the new world of America. Altizer is an American theologian who is open to the possibilities, both positive and negative, that are offered in America—and offered by America. This is a possibility for Western theology to remake itself anew, to rethink itself on what Altizer, in a letter of February 12, 2012, calls "the unique ground of America"; a society seeking to express itself through common speech and a common language. As Altizer identifies in this letter, furthermore, the death of God is in many ways centrally tied to America, a revolutionary America as expressed by William Blake in his epic *America* and as enacted in Herman Melville's novel *Moby-Dick*. This theme is a constant implicit presence in Altizer's thought and letters: what is the abyss that confronts America, that confronts American society? Why in this most religious of modern nations is the death of God so centrally, and continually enacted—and yet often denied?

Altizer's radical theology is also therefore a political theology, a radical theology that calls for a new beginning, a new beginning inaugurated by the death of God. In these letters certain names reappear: Blake, Heidegger, Nietzsche, Hegel, Milton, Levinas, and throughout the thought of D. G. Leahy. Altizer's theology is positioned in an ongoing conversation, debate, and argument with all of these names. So there is a conversation in these letters not only with those receiving the letters but also, crucially, with those thinkers of whom he writes within the letters. The net of Altizer's mind and knowledge is cast far and wide: looking back to the ancient Greeks, engaging with what can be termed radical Catholicism, and discussing ballet, theatre, literature, culture, and politics in the broadest sense. These letters are the expression and example of what theology could and should be, a theology that is inclusive in its engagement with the expressions of human life, thought, and culture. For in these letters Altizer reminds us that theology, especially radical theology, is nothing less than a continual reflexive and critical yet celebratory engagement with all of life and its possibilities. Nothing is outside the scope of theology and theological discussion. But also, in these letters, Altizer provides a crucial reminder that to attempt to do theology, to attempt to think and write theologically, to attempt to enter an understanding of modern life through the death of God, demands a deep and wide engagement with the intellectual and cultural expressions of modern life, with all that has contributed to it. This may seem an impossibility; yet, as Altizer eloquently demonstrates, theology is centrally a task of reading, thinking, and writing—ever writing one's thinking out in new ways. Furthermore, theology is undertaken in a conversation, in an engagement with others who are likewise grappling with these questions. It involves questions far more than answers, critique far more than assurance, an honesty and emotion too often kept hidden and discouraged within the walls of institutions.

Felix Pryor, in his introduction to *The Faber Book of Letters,* observed: "Those who are able to write good books are usually the ones best able to write good letters. They are at home in the medium."[4] Thomas Altizer is certainly at home in these media: in the medium of books and articles, and here, in the medium of letters. For the past 20 years he has made assiduous use of the possibilities of the internet to circulate e-mails in which his thought has been expressed in a manner that is conversational, prophetic, inquiring, and often homiletic. He has written over 300 letters to friends and colleagues in a series of meditative essays and mini-essays on religious, theological, political, and philosophical matters that are central and vital to our contemporary era. It is from this wider body that this smaller, representative collection has been selected.

These letters exemplify new possibilities for engaging in and with radical theology. Written for the internet age, they have been circulated among a broad circle of radical thinkers across North America, Europe, and the South Pacific, and so serve as the ongoing ground for a series of conversations. In this regard they are an innovative medium for doing theology in communicative dialogue: they model a deliberately inclusive manner of sharing radical thought. It has often been thought that the internet would end the age of the letter. These letters, of and for a wide generation of scholars, demonstrate the exciting possibilities for global epistles that the internet enables. It was once thought, similarly, that the decline of the letter would be a consequence of the rise of the telephone, and of course there are many conversations and discussions that occur on the telephone that once would have been put down on the page and circulated back and forth.

Yet the internet has seen a return to the letter in a way that transcends both these older formats of communication. As Brian Schroeder, one of Altizer's correspondents, commented, these letters remind "a generation that there is a dying world of intellectual correspondence that can be as important as published writings for thinking through critical questions."[5] In this Schroeder identifies a central element of these letters: they reveal Altizer thinking though critical issues in communication with a fellowship of friends and likeminded scholars. These are proclamations of intent, insights and questions; discursive essays on issues that Altizer has been thinking his way through. They range in length from essays thousands of words long to far briefer ones of hundreds of words that raise a point of immediate interest and critical insight. Above all they are communications as the thinking through of critical questions. In these we are introduced again—and in a new format—to one of the formidable intellects of the last century of thought and theology. Altizer was and remains extraordinarily widely and deeply read. He is a scholar and an intellectual engaged in a world of words and ideas, ranging freely from the past into the present, from the present applied critically to the past.

It is important to realise that in the main this is not correspondence as traditionally understood in the age of the letter addressed to a single recipient. Even in the past, the handwritten (or typewritten) letter, while addressed to a singular recipient, would oftentimes be circulated to a wider circle of acquaintances—either whole or as partially quoted in subsequent letters to others. So letters have often operated as a medium to proclaim a message to a wider network of recipients.

The roots of this are evident to all who are conversant with the biblical tradition. The history of Christianity in particular is replete with letters—from those canonized in the New Testament, to those of the church fathers, to those of theologians both orthodox and heterodox, to those of reformers and heretics. The letter has a central role in Christianity from the advent of modernity in the Reformation. The printing press enabled the letter as proclamation and call for radical reform to be widely circulated. There are also the epistles of pastoral and institutional authority.

These letters of Altizer therefore arise in the context of a deep history of the theological letter, the letter as theological document and expression, the letter as theological conversation and act of inclusion across a dispersed community. They offer a theology of immediacy and reflection, of both the time in which they are written and for our rereading again, anew. They set forth theology as first and foremost the act of communication, arising from a desire to exchange ideas, to engage with the recipients the topical questions and insights, the proclamation and prophetical impulse that abide at the heart of Altizer's life and work. This collection provides a new form of theology, the theology of e-letters, theology as internet epistle that circulates across time and space in an inclusive manner, drawing together a new community, truly a theology of transfiguration.

A Note on the Text

The letters were selected from the body of over 300 letters generously made available by Altizer and various of his correspondents. In considering how to proceed I had to decide between making thematic selections and grouping these in separate chapters or topic headings or undertaking a selection that allowed the central narrative flow of Altizer's thought to be expressed. I chose the latter course and so the letters selected occur in chronological order. This I believe preserves the integrity of Altizer's undertaking in writing these letters. They arose as expressions of his thought over almost 20 years. In reading these in chronological order we gain a far more honest engagement with how and why Altizer wrote the letters than we would if they were grouped in themes that disturbed the chronological order of composition.

Of those that I considered including, I contacted all the correspondents in question requesting permission to publish the letter. All but one agreed. The final selection is mine and mine alone; in this I sought to provide an accessible and wide-ranging coverage of the ideas and discussions included in the larger body of work. The footnotes are also my work, undertaken to provide further information for those who might not be familiar with Altizer's work and the names, events, and ideas in these letters. In working on the footnotes I was struck again at the breadth of Altizer's theology. Some names I have left without footnotes, believing that those reading this collection will be at least familiar with such figures and the broad nature of their thought.

In completing this work I thank Lissa McCullough and Brian Schroeder for their generous encouragement and assistance. They have had the privilege of knowing and working with Thomas Altizer far longer than I have—yet they have been extremely supportive of this project and for this I want to thank them.

Finally, this collection is of course the work primarily and ultimately of Thomas Altizer. Without his writing of these letters, his willingness and generosity to allow them to be edited into this collection, this project would not have come to fruition. It has been a privilege to work with one of the most original and brilliant theological minds of the past century. My first meeting with Thomas Altizer was a central inspiration for inaugurating the Radical Theologies series. I am delighted that it has enabled his voice to be heard anew.

Letter 1

To Brian Schroeder* (October/November 1996)

Dear Brian,

I am deeply grateful for your gift of *Altared Ground*,[1] not only rejoicing in its publication, but grateful for the power of the book itself, which exceeds my expectations, but confirms my deep confidence in you. First let me express my delight not only in the title but in the titles of your chapters[2]; these give expression to both an integral and an essential movement, one which has a dramatic power, and yet like most Greek tragedy fails to reach either a resolution or an ending. I am tempted to say that nothing is more alien to the Greek mind than either true ending or true beginning, and while this is the universal pattern of eternal return, here you are bold enough to identify the eschaton as a fundamentally Greek concept. And in conclusion you state that the distinction between eschatology and apocalypse lies at the very heart of this book. What can this mean?

Let me begin with your all too paradoxical title. You can write (on p. 134) that for Levinas the face is the thing in itself, the *Altared Ground* of God, this following an earlier point that the renunciation of alterity is the continual kenotic movement of the Infinite, which humbles itself in its revelation as the trace of itself, citing Levinas to the effect that the original trace is the nakedness of a "face that faces," and in the face the Infinite is revealed in all its glory and defencelessness. Frankly I can't imagine anything further from Judaism than this; here iconoclasm is truly reversed, or is this the revelation of a "past that was never truly present," occurring in the irreducible alterity of an *anarchy*, an eschatological signification without context (p. 66)? This is just how I have always understood Levinas, so I can resonate with your question as to whether there is any historical concreteness at all in Levinas, and can fully accept Derrida's criticism of Levinas; yet nevertheless I can reverence Levinas for carrying ethical thinking to its final or eschatological conclusion, wherein it becomes so abstract that it loses all point of contact whatsoever with actuality—unless for Levinas there is only one historical actuality, and that is the Holocaust itself. Surely only the Holocaust could make possible an apprehension of violence itself as an absolute violence, and of history itself as an absolute violence, and a violence inevitably following from everything we have known as a post-Platonic thinking and consciousness.

I am tempted to say that there is nothing more missing from this book than Levinas himself, his thinking is almost always presented as a reaction—above all to Hegel—but beyond that to everything we have known as thinking and history, and one enormous power of this is to unveil the illusion of everything which we have known as the ethical. Here you rightly correlate Levinas and Nietzsche, but in this domain Nietzsche is a real figure and thinker, whereas Levinas and his thinking never become concretely or actually meaningful. Even his Jewish identity is veiled, and while we do gain some concrete sense of his early reaction to Heidegger, and a far fuller sense of his reaction to Hegel, he nevertheless remains a kind of surd, and perhaps his thinking is ultimately a surd, an ultimately voiceless speech, or a contemporary *sunya* or void. Is that what most deeply attracts you? Now if the absolutely other is the other person, and if the distinction between the absolutely other and the other person collapses in the ethical significance of the face, is not that significance an eschatological significance without a context, and therefore one calling forth face itself as a voiceless presence, and one visible only in its absolute invisibility or absolute absence? Yes, this is a Derridean criticism, but now I find that inescapable, and just as everything that Levinas speaks of as face has always been invisible to me, I now sense that this is absolute necessary and inevitable.

Now you already know how deeply I disagree with your political interpretation of Hegel, but let us examine its apocalyptic form, as when you state that the apocalyptic vision of Spirit rests on the establishment of a universal and total State (p. 70). Let us leave aside the question of whether such an interpretation could be supported by even one text of Hegel's and inquire into its understanding of apocalyptic, since you declare that the distinction between eschatology and apocalypse is so central here. If we follow Levinas and understand eschatology as a relation with a surplus always exterior to the totality, or to an infinity transcending totality, nonencompassable within a totality and yet as primordial as totality, and yet as a "beyond" of history drawing us out of the jurisdiction of history, and doing so by an ultimate judgment calling forth our full responsibility, then there is certainly nothing Greek in such an eschatology, nor anything truly messianic either; rather, this understanding of eschatology is characteristic of the pre-exilic prophetic oracles. For here Levinas does take seriously the ancient prophets as you do not, and if we follow Levinas here, would we then understand apocalypse as totality itself, or as historical totality, or as an absolutely immanent transcendence, or as the total presence of the "Kingdom of God"?

Thereby, as you know, I understand both Hegel and Nietzsche as purely and absolutely apocalyptic thinkers, but I do not here see how you understand apocalyptic. One way to do this, which you never employ, is to understand *eschatological* in this sense as Jewish or Judaic and *apocalyptic* as Christian, thereby allowing us to see that there is a deep conflict between Judaism and Christianity. I recognize that this is a tactic that you must refuse, but then I remain baffled by your understanding of apocalypse, unless you want to understand it as being purely Hegelian. This takes me back to your "BackGround," where you state that the predominant tendency in Western thinking has been to construe the Absolute or Ultimate either as illusion, nothingness, or phantasm, or as radical exteriority. Now if Hegel is the former, a truly postmodern interpretation, and Levinas is the latter, is the former apocalyptic and the latter eschatological? While I missed this in "HyperGround," perhaps it was there in your assertion that the task confronting contemporary thinking is the formulation of new nonfoundationalist conceptions of ground and Godhead, where the challenge is the determination of the Absolute either as one that allows for the possibility of a genuine once-and-for-all beginning that would be synonymous with the fullness of knowledge, grace, and salvation, or as the absolutely other that conditions all knowledge and ethics on the very basis of its ineffable transcendence and radical alterity. The former apocalyptic and the latter eschatological? Is Levinas's interpretation of the metaphysical One as ethical social multiplicity, as a nontotalizing thought, which is offered here as the decisive insight of twentieth-century philosophy, an eschatological One which is absolutely other than "the concept of apocalypse which dominates classic metaphysics or ontotheology" (p. 141)? Is this that One calling forth a discourse with the Other as discourse of God, and does this alone make possible the true task ahead for thinking and action, which is the "continual production of divinity" (p. 147)?

Admittedly this is a cryptic conclusion, or perhaps no conclusion or ending at all, but is that necessitated by an eschatological ground, and an eschatological ground that can only be real as a totally nonapocalyptic ground? But is Levinas's thinking a continual production of divinity? Or is "production" only an eschatological witness and thinking, and one made possible only by a purely pagan apocalypse or totality, an apocalypse that in our time has become a purely nihilistic apocalypse, but perhaps precisely thereby is now calling for the deepest possible eschatological witness and thinking?

Yours with deep gratitude,
Tom

P.S. Will you visit me this summer? I hope so!

Letter 2

To Brian Schroeder (December 5, 1996)

Dear Brian,

I was extremely impressed with your last letter responding to my response to *Altared Ground*. Yes, you do have a genuine theological mind, which is extremely rare today, and perhaps one crucial ground of your academic difficulties. Let me begin this response by speaking of a crucial historical difference that lies between us, and that is whether or not the Greek philosophical mind, and more specifically Plato, truly apprehended what we have subsequently known as the transcendence of God. I join the common position in seeing this as perhaps the greatest difference between Plato and Neoplatonism and in maintaining that transcendence is alien to classical Greek thinking, and even perhaps above all absent from the later Plato. Now as far as I know, and admittedly I am out of touch, there is no scholarly exegesis which finds such transcendence in Plato, with the exception of Eric Voegelin,[1] whom I thought stood alone here, just as he did in his great project of establishing a full coincidence between Athens and Jerusalem. While Platonic thinking is certainly grounded in transcendence, this is a transcendence in some fundamental continuity with the realm of becoming, and not a transcendence that is pure or absolute transcendence alone. I also join the common position in believing that an absolute transcendence first dawns in Israel, but not until the prophetic revolution of the eighth century BCE, so that it is indeed absent from the earlier traditions of Israel that you mention. Thus I believe that an absolute or pure transcendence in its origin is uniquely biblical, having no true or full parallels elsewhere in the world, and this is the transcendence which is the deepest ground of Judaism, Christianity, and Islam, and inevitably the incorporation of Greek philosophical thinking into these traditions evoked ultimate controversies, but it is in the Catholic and the Orthodox Christian traditions alone that Greek thinking was absorbed into orthodoxy, and as a Lutheran you know full well what an ultimate offense such orthodoxy is to a biblical Christian.

I feel embarrassed in saying these things, of only because they are so conventional, but I simply do not know where they have been truly challenged, except, of course, by catholic thinkers. Does Levinas challenge this position? Where? And does Levinas believe that eschaton is a fundamentally Greek concept? Where does he show that it is Greek thinking that apprehends the primordial radicality of transcendence? And what is the relationship between a primordial transcendence and the eschaton? Eastern Christian thinking can know the apocalypse as a return to a primordial eternity or transcendence; is Levinas within such a paradigm of eternal return, and is this true of you too? I cannot believe it, and cannot do so if only because you are deeply Nietzschean, and thus committed to an eternal recurrence that is the very opposite of eternal return. And how is the eschatological dimension one of height? Does eschatological for you simply mean transcendent? Is it fundamentally a spatial rather than a temporal category? This is surely an odd employment of the word *eschatological*, and it will inevitably create confusion. But if *eschatological* for you and for Levinas is most deeply the assumption of the infinite responsibility imposed by the "other" on the "same"—one calling humanity out of the jurisdiction of history or the "totality"—surely such responsibility is nowhere found in Plato, or even within that Christian thinking that has fully absorbed Greek philosophical thinking. Even Leo Strauss[2] finds a polarity between Athens and Jerusalem here, and I find it extraordinarily difficult to believe that Levinas could think that there is a Greek philosophical ground for what he calls forth as the Infinite. If so, why is an Infinite so fully absent from ontotheology? Or so fully absent from that Heideggerian thinking which intends a radical deconstruction of ontotheology? Here, I think, lies the deepest difference between Heidegger and Derrida, a difference that Derrida philosophically learned from Levinas.

As I have said many times, a deep difference I sense between my generation and yours is what I apprehend as an absence of a historical consciousness in younger thinkers, and even in many who are only a bit younger than I am, such as Bob Scharlemann.[3] True, such an absence can make possible a new thinking, one that you promise in understanding the "face" as the locus of the trace of the infinitely other, and if the face is the *altared ground* of God, then does this truly make possible the meaningfulness of the nontheoretical eschatology of the will to power? If so, I think that you will be forced to belie or transcend any language about subject or subjectivity, or else you will wholly lose a Nietzschean ground. Now if the paradox of eschatology is the nonsimultaneous incommensurability that appears in the face, a face that at bottom is the will to power in a Nietzschean sense, this is surely one way in which to apprehend the face apart from the subject, or to know the face as the very "other" of the subject, an other that is truly a trace of the infinitely other. Yet I sense a Pauline ground here, one knowing and even first calling forth a pure dichotomy deeply within the subject or the "I," a dichotomy between *sarx* and *pneuma*,[4] or between sin and grace, or even between old aeon and new aeon. This is apocalyptic—and yes it is Hegelian, too, so I presume that you would resist it—but is it possible either to know or to apprehend a nondichotomous or nondialectical infinite responsibility?

I sense that you are most deeply called to ethical thinking, one surely present in Levinas, and perhaps in Levinas alone among truly contemporary thinkers. This I presume is why Levinas is making such a deep impact today, but you are surely rare in understanding how this demands a radical transformation of our ethical thinking, one unveiling the profoundly nonethical ground of all our given ethical thinking, and one simultaneously unveiling the profound necessity for a genuinely theological ground for ethical thinking itself. But surely a theological ground demands a calling forth of God, one which Levinas himself surely does, yet is this possible for you? And possible for you as a genuine Nietzschean, or a genuine witness to a uniquely Nietzschean will to power? Is it possible for you to understand the Will to Power as the absolute immanence of God, an absolute immanence wholly reversing everything that we have *known* as transcendence? Could the true eschatological be an absolute transcendence of transcendence, one which is simultaneously an absolute transcendence of immanence, too, so that if totality is either a pure transcendence or a pure immanence, the truly eschatological is a transcendence of every possible totality, hence it is the death of God and the death of the subject at once, but only that simultaneous death makes possible an infinite responsibility.

This I believe is the deepest ethical question, the very meaning of responsibility, or the very possibility of responsibility, and above all for us the very relation of responsibility to totality. If you must refuse every possible Hegelian *Aufhebung*[5], or every negation which truly preserves what it negates, is responsibility as such, or an infinite responsibility, a negation of totality wholly shattering every possible totality, or inevitably dissolving totality in a moment of infinite responsibility? This I think is deeply Nietzschean, but it is a responsibility or a Yes-saying that is impossible apart from the death of God, or the death of transcendence itself, or at the very least the death of every transcendence that our history has known.

Such a responsibility could be understood as nihilism, which you seem to do when you accept Nietzsche's affirmation of the necessary inevitability of nihilism to be the first truly modern affirmation of the eschaton. Here you are apparently deeply distant from Levinas, but thereby contemporary in a way that Levinas cannot be, and perhaps this is a genuine way to a truly contemporary ethical thinking, calling forth ethical thinking as nihilistic thinking, and nihilistic if only because it demands a negation of totality, and above all a negation of the totality of a uniquely contemporary postmodernity. If there truly is a postmodern thinker, that could only be Nietzsche, unless it is most recently D. G. Leahy, and this is a thinking most manifestly or most clearly negating all possible ethical thinking, or every thinking in which there is a subject of the ethical act, or in which there is any subject or subjectivity whatsoever. Here, the ancient Greeks can exercise a spell upon us as they did upon Hegel, Nietzsche, and Heidegger, for they were innocent of what we have known as the subject, a subject which is even absent from the highest moments of Greek sculpture, and even from Greek tragedy with the possible exception of Euripides, just as Greek philosophy is innocent of the pure subject of thinking, a pure subject which was only discovered by Descartes. Perhaps such a pure subject is what most deeply makes eschatology as you understand it impossible, so that Nietzsche is our first modern eschatological thinker, even if this is a thinking ending modernity itself, and ending it by dissolving the subject of thinking. But can we also understand Nietzsche as our only purely eschatological thinker? Is that, indeed, how you understand him?

Tom

Letter 3

To D. G. Leahy*
(January 17, 1997)

Dear David,

In response to your letter of January 10, I find that my earlier concerns are only deepened. Even if your letter is genuinely enlightening, nevertheless I am still not able to understand what you might possibly mean by otherness, and the genuineness of otherness, simply by the affirmation that it is not compatible with either "ownness" or "intrinsicness." If the Very First absolutely displaces intrinsicality, so that intrinsicality is absolutely ended, how could otherness of any kind then be possible? This only deepens the problem of evil, or dissolves it altogether. Yet I gather from your few comments on evil that you confine it to a history that is already ended, for even if you say that evil is quite real world-historically, and that a modern self–other dichotomy is itself an embodiment of evil, then if such modernity and such history have come to an end, is not evil itself now wholly unreal? Now it is true that this is a deep problem for all apocalypticism, or for all full apocalypticism, and I do not see how you can escape it, and particularly so if you refuse to become open to the problem or the mystery of evil in the Godhead. Once again the problem of what you mean by an absolute nothingness comes forward, and if you will not abandon the horizon of scholasticism, you must do far more with scholasticism's identification of evil and nothingness.

Indeed, in this letter you do something which is new to me, and extraordinarily precarious, in identifying the absolutely new language with the language of I AM and the language of Jesus. I believe that this is one of your most vulnerable points, and particularly so insofar as you refuse modern critical and historical language as a language that has come to an end. True, this occurs in a purely Christian mystical language, but that, too, is a language that you refuse, even as you also refuse anything that is manifest as an imaginative language. Now if essential continuity with all previous forms of thought or species of language is precisely what is denied here, how can you claim a continuity with the language of the Bible, unless this is understood in something like a purely mystical sense? Once again the deep problem of the relation between apocalyptic and mystical language, or primordial and apocalyptic language, arises, and if you are claiming that an absolutely new language coincides with biblical language, or the deepest or purest biblical language, is this a language that is wholly silent and unheard apart from the full and final advent of the Very First? If so, what then happens to Catholic tradition?

This problem is only deepened by what you say about the language of Jesus. Yes, I think we can say that this is an absolutely common and an absolutely new language, one that I can thereby only understand as a *coincidentia oppositorum*,[1] but in refusing any such possibility, how can you be open to such language, or does your understanding of the language of Jesus have no continuity whatsoever with *anything* that previously was understood as the language of Jesus? For example, must you wholly distance your understanding from all biblical scholarship? Is everything that we have known as biblical scholarship and biblical exegesis now absolutely ended? Is all such understanding the expression of a subject or a self that has now ended?

Here another deep problem presents itself, one that as far as I know you have not dealt with, but one that I believe is absolutely essential to your thinking. For if subject or selfhood has now ended, is it possible to understand that ending without understanding the origin or beginning of selfhood and subject? Let Nietzsche once again stand as a model, for he fully conjoins an understanding of the ending of subject or *ressentiment*[2] with an understanding of its beginning, and it is precisely his understanding of the origin of repression or the "bad conscience" that makes possible his understanding of its ending. In a far more complex and comprehensive form, this is true of Hegel, too, and if Hegel understands the Incarnation as the absolute and final inauguration of the full actuality of self-consciousness, a self-consciousness that is finally and only absolute Spirit, this is a self-consciousness that precisely thereby is a kenotic consciousness, one undergoing an ultimate death or self-negation, a kenotic self-negation that alone makes possible both its origin and its fulfillment or consummation. Yet despite your ultimate centering upon absolute beginning, an absolute beginning that is an absolute ending, your thinking appears to be wholly closed to an understanding of the origin or beginning of that which is now ending or has ended. Once again I am reminded of a primordial language, and most purely a Buddhist language, one in which no actual or real beginning is possible, or one in which omega is quite simply and only alpha, and if this is the language of Eastern Christianity, or of a purely mystical Christianity, is it not thereby a language that is coincident with your own?

I believe this problem is even related to your insistence that Aquinas did understand the death of God as having occurred in the Passion. For you reveal perhaps something essential about your thinking in speaking of Thomas's understanding of the Person of the Incarnate Word, for in ST [*Summa Theologica*] III, his language is most deeply and manifestly drawn from the language of the "Damascene," and if John of Damascus[3] finally became the most influential of all Eastern theologians, he was so as a deeply Neoplatonic thinker, and it is his

Neoplatonic language that Thomas most fully employs in speaking of the Person of Christ. Observe this language:

> As Damascene says, the Divine Nature is said to be incarnate because it is united to flesh personally, and not that it is changed into flesh. So likewise the flesh is said to be deified, as he also says, not by change, but by union with the Word, its natural properties still remaining, and hence it may be considered as deified, inasmuch as it becomes the flesh of the Word of God, and not that it becomes God.
>
> (ST 3.2.1)

So it is that Thomas understands the crucifixion not as the death of God, but only as the death of Christ's carnal body, for Christ is wholly divided in his human and divine natures, for while Christ is God made man, and even God "truly humanized" (ST 3.2.6), the union of the divine and human natures is a union occurring only in the "person" or the individual substance of the incarnate Christ, and not one that is truly present in his passion and death. So that if the "thinking now occurring for the first time"[4] also for the first time truly and finally thinks the death of God, this is not only not in continuity with scholasticism, but has no precedent whatsoever in scholasticism. But that raises the question of whether or not the full and pure understanding of the death of God that occurs here is an absolutely unique understanding, with no origin or history whatsoever, and if that is true, is the thinking now occurring for the first time an absolutely isolated or absolutely solitary thinking—or is it such if it is not, in fact and in actuality, the thinking of that absolute apocalypse that has already fully and finally dawned? But if so, and if the historical events of 1989[5] are a decisive sign of that advent, is it here and only here that an understanding of that apocalypse is at hand?

Tom

Letter 4

To Lissa McCullough*
(June 13, 1997)

Dear Lissa,

Once again I am employing a communication with you in an attempt to renew my writing. Let us presume that I can enter new seas by centering upon "Yes-saying," upon life rather than death, upon light rather than upon darkness. Hopefully by this time I have absorbed the Lutheran lesson that there can be no Christ of glory apart from the Christ of passion, and that it is precisely the Christ of glory and the Christ of glory alone who most reverses the gospel. It is true that we live in a time when joy or true joy is most invisible and unheard, so perhaps it is now a supreme theological task to call forth or give witness to an embodied joy, and perhaps the time has arrived to speak of the Resurrected Christ. So it is that I am now musing upon doing a book on the Resurrection, although it will have to be a book on death and resurrection, with the hope that now and for the first time I can center upon resurrection rather than upon death.

Yet I cannot escape the conviction that Gnosticism remains our supreme challenge, and insofar as it is here that we may discover our purest dissolution of death, the Gnostic threat remains inescapable. So I am tempted to begin with a chapter on the advent of death, and the ultimate advent of death, which could only be the advent or genesis of God, but I will veil that theme here, and concentrate instead upon the advent of an ultimate darkness that is the necessary arena to make possible a transfiguration that is an absolute Yes-saying or resurrection. Yes, the dialectical identity of crucifixion and resurrection remains fundamental in my thinking, but as you know all too well, thus far I have written forcefully only upon crucifixion, only upon death and darkness and not upon life and joy. If Augustine and Barth are our greatest pure theologians, nowhere is this more manifest than in their integral and even dialectical conjunction of sin and grace or damnation and salvation, and at no other point are we more deeply in need of truly new theological thinking today. Is it possible to achieve this in a new thinking about resurrection?

Now even if it is true that no classical theological motif is more muted in modern theology than is resurrection, all too significantly it is wholly absent in Schleiermacher's dogmatics, it is nevertheless true that it is a primal motif in the deeper expressions of the modern imagination, perhaps being wholly absent only in Melville[1] and Kafka,[2] just as it likewise is an ultimate and even absolute motif in the deepest expressions of modern thinking, with the inevitable exception of Kant, unless Kant's understanding of the sublime in the third critique is a modern reflection of resurrection. Perhaps it is Kant above all thinkers who impels us to face the necessity of a profound transformation of our theological language, one that no modern "theologian" has dared accept, and if the purest expressions of modern poetry, painting, and music have called forth an ultimate and final joy, even if a joy inseparable from its true opposite, such joy must be known theologically as resurrection, but precisely thereby as a resurrection from death. But the ultimacy of resurrection is inseparable from the ultimacy of death, and if it is precisely the latter that is the most pragmatic test of the theologian, perhaps the ultimate theological test is the challenge of speaking or evoking the ultimacy of life, or is so in our contemporary situation and world.

Let me at this point pass on to more mundane issues. Despite the fact that we have long since known the deep importance of distinguishing resurrection from immortality, I am aware of no clear or decisive theological distinction between them, except insofar as it is a truly historical distinction. Now that we are aware of the deep importance of language, and hopefully equally aware that the very words "resurrection" and "immortality" are truly alien to a genuinely or fully modern language, we must become open to the challenge of this void, and recognize it as a positive and not simply negative challenge, one truly realized in our fullest language, even if that language is now theologically unrecognizable. Most simply stated, the challenge is of speaking of a death that is inseparable from life, and just as we have learned that "life" is only truly speakable through a language of "death," we must become open to a "resurrection" that is inseparable from "crucifixion," and one way to do this is through a "resurrection" that is wholly other than "immortality." Yes, theologically we have long known this, but we have not been able to speak it, or not in an overt or manifest theological language, or a theological language that has any real point of contact with our common language. Theologically stated, is it possible to speak or evoke a "death" that is fully and actually "life," but is so only insofar as it wholly dissolves everything whatsoever that we can know or imagine as "immortality"? At this point something happened on our trip to the South that had an impact upon me, that was my purchase of Herbert Grierson's anthology, *And the Third Day: A Record of Hope and Fulfillment* (1948),[3] for despite the fact that this anthology contains a great many writers whom I revere, this very focusing upon immortality effects a pure nausea in me, as I am once again repulsed by a transformation of the Christ of Passion into the Christ of Glory.

Now if we follow Luther, Kierkegaard, and Nietzsche at this point, we can know such immortality as a deep and profound offense, and not only an offense but a true reversal of life and joy, and precisely thereby a reversal of the gospel itself, a gospel here truly becoming "dysangel,"[4] although this is perhaps only actually so in the modern world. Let us attend to what is most manifestly "dysangel" here, is it not a language speaking life and life alone, an "eternal life" wholly silencing "eternal death," but thereby and precisely thereby a language of death itself, a language of a death wholly silencing "life"? Thus if we can know such a language of life to be at bottom a language of death, is it possible to reverse such language in speaking of resurrection, as Blake[5] and Nietzsche in their purest language surely did, and not simply to reverse it in our theological intention, but in the actuality of our language itself? Here, I am impelled to return to that New Testament language that I most revere, the Gospel of Mark and the genuine letters of Paul, attempting to become attentive to how these languages deconstruct or demythologize themselves, for then a gospel emerges that is the very opposite of anything that we can know as the kerygma of the church, as "resurrection" is the very opposite of "immortality," and is so insofar as it wholly shatters our deepest longing, or our deepest innocence.

So to return to my initial problem, how do I begin? But how does the Gospel of Mark begin? For it begins with the beginning of the gospel of Jesus Christ, thereby not only is it distanced from the gospels of Mathew and Luke, who can only know beginning as an eternal genesis or genealogy, even if it is here at one with the Fourth Gospel, and I have long been intrigued by the seeming if as yet uncomprehended parallels between the gospels of Mark and John. Certainly remarkable parallels occur in their enactments or renewals of the resurrection, for even if Mark minimizes the resurrection in his narrative, John, unlike Mathew and Luke, but paralleling Paul, wholly integrates the resurrection with the crucifixion, so that in neither John nor Mark, as opposed to Mathew and Luke, is the resurrection in deep continuity with the Hellenistic mystery deities and with all Gnostic visions of resurrection. Perhaps this is the very point at which the Fourth Gospel is most "realistic" in Auerbach's sense,[6] and even if this was a consequence of that profound conflict over Gnosticism that so deeply divided the Johannine community, the first epistle to the Corinthians is decisive testimony as to how a revulsion against Gnosticism led Paul to center upon the crucifixion, and if this epistle concludes with a vision of resurrection and of resurrection alone, that is a vision that is virtually inseparable from Gnosticism, and one also anticipating that profound and overwhelming belief in immortality that so dominated both the ancient church and Christendom itself.

So can I begin with a death that is a reflection of the gospel, and even a reflection of an authentic language of resurrection, a language of death that is finally inseparable from a language of life and joy? If I choose to begin with a language of the advent of death, can such an advent reflect the beginning of the gospel, a beginning that is an actual beginning, and not only an actual beginning but an ultimately actual beginning, a beginning that is absolute genesis? I see this as the challenge here, for if the gospel is a gospel of absolute genesis, is that a genesis that can be isolated from a language of death, and above all so if the advent of death is an ultimately actual beginning. The truth is that just as resurrection is inseparable from death, redemption is inseparable from fall, and an ultimate redemption inseparable from an ultimate fall. No one knew this more deeply than did Blake and Nietzsche, and I think that this is also true of Paul and the Fourth Gospel, even as it is true of Isaiah and Job, and surely the *Book of Job* is that scripture which is most closed to resurrection, even if its desperate assault upon the Almighty or El Shaddai issues in a cry for vindication against the Almighty (*Job*19:25), a cry that the orthodox all too ironically understand as a vision of the resurrection. Yet is there not a hidden truth here, and a dialectical one at that, one resurrected in our own world by Kafka and Beckett,[7] for an ultimate speaking in response to an ultimate darkness is finally a transfiguration of that darkness, and one impossible apart from its own self-lacerating modulations. If it is possible to know the advent of death, and even know it in our deepest responses to death, must not an ultimate advent be gospel rather than *dysangel*, and gospel precisely in calling forth an absolute genesis?

Seemingly, we true moderns can only know grace as apocalypse, and an absolute apocalypse as absolute grace, but such a grace is known by both Augustine and Barth as occurring in genesis, just as the established world of theology can only know apocalypse as a repetition or renewal of genesis. Yet if that theology can truly be reversed, and reversed above all in its absolute refusal of the absolutely new, is not such a reversal finally Yes-saying rather than No-saying, and Yes-saying in so greeting the absolutely new? Here, lies a decisive clue that I am seeking, an integral association if not an identity between *novum* and affirmation, and between absolute affirmation and the absolute *novum*. Is the advent of death a reflection of absolute *novum*? Is a true and actual advent, and above all an absolutely actual advent, an inevitable consequence of absolute *novum* itself? Innumerable critics have sought a true parallel between modern totalitarianism and ancient Gnosticism, but is not one such parallel their absolute reversal of the history and world of their own time, wherein what presents itself as an ultimate call to the absolutely new is at bottom a call of return to an absolutely primordial world, a world innocent of everything we have truly known as history and humanity, and innocent because it is a primordial or absolute beginning? Now such a beginning is the absolute "other" of an absolute genesis that is the genesis of the absolutely new, and perhaps most "other" in its ultimate sanctioning of an original nature or world, a world in which no true advent is possible, and thus a world that is finally a cosmic eternal return.

As I muse upon these matters, I am coming to recognize that the crucial word here is "advent" rather than "death," and if Advent is the advent of the Passion and the Resurrection of Christ, and if this is the advent of the absolutely new or of absolute *novum*, it is thereby the advent of absolute genesis, and of that absolute genesis that is absolute event or absolute actuality. Finally, we can know that advent only as the advent of Joy, or of absolute Body, or of that body which is the Body of Resurrection, or of that body releasing an absolute and final Yes-saying. Yet it is of overwhelming importance that we recognize this body as an absolutely new body, it could only be the very opposite of everything whatsoever that we can know or imagine as primordial body, and if, as I am persuaded, the very advent of this body immediately called forth its reversal in Gnosticism, and a Gnosticism that was absorbed into the dominant body of the Church and Christendom, this is a Gnosticism that is a renewal of primordial body, just as it is a renewal of primordial return, and is so most clearly in so deeply knowing and celebrating a primordial "eternal now." This is the eternal present that is the absolute opposite of the absolutely new, and therefore one foreclosing any possibility of true or genuine advent, so that the eternity that it knows could only be a primordial or original eternity, but therewith the very reversal of a new eternity or a new aeon.

Western theology, as opposed to Eastern theology, has refused to open itself to the historical truth that this eternity is truly new, truly new historically, going far beyond anything that the Classical or Hellenistic worlds could know as eternity or Godhead, and doing so by realizing an absolute transcendence that is historically new, but one that could only be an absolutely primordial transcendence, and one that is the consequence of an absolute reversal of the absolutely new. Now immortality is the center of a new life as it had never been before, and now a truly new longing for death is released in the world, one with no real historical precedents whatsoever, just as an "other-worldly" spirituality is now triumphant, even if it is finally inseparable from an all too human will-to-power, one that in the Constantinian era established the most politically powerful religious world in history, yet a religious world that was more "other-worldly" than any previous religious world. Why do we theologically refuse to take this profound conjunction seriously? Or is that yet another deep ground of the radical Christian from St. Francis and Meister Eckhart[8] to Blake and Nietzsche and Weil[9]?

Is it not true advent itself that is most alien to our established religious world? Is it not possible that at least for us it is precisely by truly and fully speaking of advent that we most clearly give witness to grace? And could it be that it is precisely by speaking of the advent, and the genuine and full advent, of death and darkness that we can be open to the transfiguration of darkness? Once again Kierkegaard's crucial distinction between recollection and repetition is deeply important, the one a backward and the other a forward movement, or the one a primordial and the other an apocalyptic movement, and if this is the distinction that most openly calls forth the chasm between paganism and Christianity, always remembering that for Kierkegaard "paganism" is Christendom, then it is recollection or eternal return that forecloses the possibility of "advent," just as it is repetition or eternal recurrence that is true "advent." Thus the advent of death is gospel, and is gospel precisely because it is advent, for advent itself is "resurrection," and resurrection precisely because it is absolutely new. Therewith we can clearly see how immortality is the very opposite of resurrection, and not opposite in a dialectical sense, but a final or closed opposite precluding the very possibility of a genuinely dialectical opposite. So, too, contemporaneity with Christ in a Kierkegaardian sense cannot be contemporaneity with the primordial Christ, or with the Christ of Glory, but only contemporaneity with the Christ of Passion, or with that Christ who is the absolutely new Christ.

Tom

Letter 5

To Brian Schroeder (January 18, 1998)

Dear Brian,

I much appreciated your response to my "rejection," and yes, I do know that theology is coming to a close, even if I cannot accept it. But the question of philosophy and theology today is a much more difficult one, in large measure because the theological dimension of philosophy since Nietzsche has been so profoundly elusive, as is most manifest in Heidegger himself. Frankly I think that Heidegger's thinking is both Christian and pagan, with the Christian ground manifest in *Being and Time*, but apparently progressively disappearing after the *Kehre*.[1] And just as we still don't understand the role of Nazism in that turning, the theological meaning of the later Heidegger is extraordinarily difficult, and I am wrestling with that now in reading Reiner Schürmann's excellent book on Heidegger.[2] Certainly you are right in speaking of the revulsion of philosophers against Christian dogmatic categories, and this is even echoed in many theologians today. I think that an effective pragmatic approach to this problem is to approach it historically, noting how profoundly theological all major philosophy is until the advent of the modern world, and one of the things I liked most about Ed's new book[3] is the role he gave theology in the evolution of the understanding of space, demonstrating that it simply cannot be understood apart from theology. Of course, this is not dogmatic theology, but I would maintain that the contemporary meaning of *dogmatic* itself did not evolve until modernity, and that this reflects a deep gulf between theology and philosophy that did not arise until the birth of the modern world, as perhaps most decisively present in Luther himself. But if we are now truly realizing the end of modernity, and therefore the ending of modern philosophy, then there is at the very least the deep potential in this for a truly new relationship between philosophy and theology, and certainly Leahy's "thinking now occurring for the first time" does profoundly exemplify this.

I remain shocked and offended that contemporary philosophers ignore Leahy, but so too do almost all contemporary theologians. Another and perhaps related problem is present, and that is that philosophers and others today almost universally associate the theological itself with the deeply conservative or "orthodox," seeing it as a deeply regressive and pathological movement. Of course, this is justified by the very state of theology today, but I think that this critical situation is deepened when philosophers themselves do incorporate theological thinking in our time, but only so incorporate conservative theological thinking, as witness Gadamer, Ricoeur, and even Levinas. At this point let me say that I do not think *The Symbolism of Evil*[4] is a genuinely philosophical work, although it is genuinely hermeneutical, even if it contains little that is new to the religious scholar. And I am intrigued to learn about Levinas's writing about evil, even if it is only in one essay,[5] and if you can get your departmental office to send me a copy please do so, although I found the last Levinas papers you sent me virtually unreadable since they were so naive theologically. So this I see as a deep problem, that theology only has a deeply conservative identity upon the contemporary philosophical horizon, even if this was not true as recently as the time of Whitehead and Wittgenstein (incidentally the big biography of the latter intrigued me in its portrait of Wittgenstein's deeply Christian life).[6] Of course, this is another world today, as witness Husserl's letter of recommendation for Heidegger, in which both the writer and his subject are portrayed in strongly Christian terms.[7]

Another way of approaching this problem is to take due note of the profound role of religion in late modern literature, art, and music, and never is this dogmatic in the common sense; if anything, here "dogma" is commonly and truly reversed. Yet this occurs since the very advent of modern literature in Dante, and ever more gradually and progressively deepens and expands, so that our fullest modern literature is deeply and comprehensively "profane," and yet even thereby profoundly inseparable from the "sacred." Now this is true of Nietzsche, too, and of Spinoza and Hegel as well, so there might be some such coincidence between twentieth-century literature and philosophy, even if this is invisible to the professional philosopher. Once again we are victims of our contemporary academic divisions and specializations, and yet these are clearly blocking deeper inquiries, and just as they are being discarded in a new science, they should also be wholly transcended in a new philosophy. I deem it amazing that philosophers who truly want to be "relevant" in America today can proceed as though religion and theology are of no significance at all, this would appear to be both indifference toward and contempt of the clear majority of the American people, to say nothing of the great majority of people throughout the world as a whole. The truth is that few intellectuals today beyond philosophy departments take contemporary philosophy with any seriousness at all, and I would think that one of the first things that philosophy must now transcend is its purely academic identity. And come to think of it, few truly major philosophers have been academics in our sense, and apparently nothing so grieved Wittgenstein as did his academic life, a life making genuine thinking impossible.

Incidentally, I have read Derrida's *The Gift of Death,*[8] think very little of it, and am sad that this can now be thought of as theological thinking. Not that I don't genuinely respect Derrida as a theological thinker, which I think of as a fundamental core of his real work, but I think that he is just as superficial here as in his other recent writing. I think that the fire has gone out of Derrida, and although this commonly and almost universally occurs as we age, it does not always so occur, as witness Ed.[9] Now as you know I do have one deep and unusual conviction about philosophy, and that is that it is not truly important or real unless it is theological, I think this is clearly manifest in most of our great philosophers, and that its absence from philosophical thinking today is a decisive sign of the crisis of philosophy. You will remember that I showed you Cornelio Fabro's *God in Exile*[10] when you were last here, which if nothing else is a superb history of modern philosophy, but it also fully demonstrates how a reverse or inverted theological thinking is so absolutely fundamental in that philosophy, and just as his enactment of it is an epic enactment, it fully illustrates how a theological understanding of that philosophy can deeply illuminate its truly organic and deeply consistent development and coherence. And let us also remember that Hegel's lectures on the history of philosophy give a very prominent role to its corresponding theological development, and do so by effectively suspending or ignoring "dogmatic" theology.

Perhaps that is what we most need today both philosophically and theologically, a true divorce between "dogmatic" theology and a deeper or more modern or postmodern theology. Yes, I am "dogmatic," but am so only in a real attempt to invert and reverse dogmatic theology. I do believe that this occurred in Spinoza, Hegel, and Nietzsche, and perhaps in every major modern philosopher. But something like this occurred in ancient and even medieval philosophy as well; even Augustine can arouse deep fury in the Eastern Christian theologian, and a contemporary orthodox theological thinker like Marion can deeply attack Aquinas, just as orthodox medieval and even modern Jewish theologians have furiously attacked Maimonides. For nothing is a deeper threat to faith than thinking itself, unless this is true of the imagination as well, and if here our deeper imagination and our deeper thinking are truly conjoined. Can either one truly be itself without such a relation to "faith," or without a truly theological ground? Yes, we can understand that as a negative ground, and even a purely negative ground, but it is theology which first called forth such a negative ground, and did so in its very birth in Paul, and just as I understand both Hegel and Nietzsche as deeply Pauline thinkers, although reverse ones, it is in Paul that a pure subject or "I" is first manifest, that very subject which realizes its philosophical birth first in Augustine and then in Descartes, and a subject whose echoes even Heidegger must deeply contend with, for that is a "subject" inseparable from a theological ground and that can only perish through what we know as the death of God.

Yours faithfully,
Thomas J. J. Altizer

Letter 6

To Brian Schroeder (January 28, 1998)

Dear Brian,

Thank you very much for sending me the Levinas papers, which I have now read. These are intriguing, even if they say little about evil, although that little may be important. The essay on "Transcendence and Evil" is as you know largely a response to Nemo's book on the Book of Job,[1] an essay disquieting to the theologian if only because it so ignores all critical historical understanding of the Book of Job, but more deeply because it ignores the deeper theological problems posed by the book, problems revolving about the purely negative portrait here of El Shaddai. Is Levinas wholly closed to this problem? Once again Levinas reveals his deep move to the purely primordial, but I am most intrigued here by his evocation of a nothingness which would not only be the nothingness of negation, but a truly radical alterity, for this may well be a genuine response to the Book of Job.

Now, if Levinas is inspired by Husserl toward a purely phenomenological quest for originating meaning, could this be embodied in an anxiety that is "the cutting point at the heart of evil"?[2] Anxiety as the root of all social miseries and human dereliction? Is this only Nemo or Levinas too? Then Levinas appears to speak for himself in declaring that evil will indeed mean the "end" of the world, and an end leading beyond—beyond both nothingness and being. Here, in my judgment, lies the heart of the matter. But does Levinas explore it? Does he do so in speaking of the "original phenomenality" or quality of evil, a "counter-nature," and in that sense transcendence? Surely he does not do so in speaking of evil as an excess, but note that this is the context in which he speaks against theodicy, identifying theodicy as a way of conceiving God as a reality of the world. This *is* a response to the conclusion of the Book of Job, and I could accept it as that, even if in my judgment this is a conclusion deconstructing the very heart and substance of the book (Job's dialogues with his friends).

He can declare that the original "intentionality" of the relation between beings is a relationship with God, and can go so far as to affirm that the first metaphysical question is no longer "why is there something rather than nothing," but rather "why is there evil rather than good?" And this he can identify as the de-neutralization of being, or the "beyond being." Here, meaning itself can be understood to begin with the soul's awakening by evil—"God does evil to me to tear me out of the world"—and only now do I become unique and exceptional as a soul. The *felix culpa* indeed![3] Once again this leads to the priority of the ethical over the ontological, and a priority only possible in a relationship with the "you" of God that *eclipses* being. Now dialectics is wholly transcended, there is no possibility of knowing evil as any kind of negation, it signifies only "ex-cess," and this is an excess refusing any synthesis in which the "wholly-otherness" of God would become visible. So it is that the theophany of the conclusion of the Book of Job is the theophany of the absolutely Wholly Other, and presumably this occurs through the soul's being awakened to the horror of evil, yet this horror is a breakthrough of the Good which is not a simple inversion of Evil, but rather an "elevation." Now we can truly know that the Good does not please but commands and prescribes, and obedience to prescription is the very elevation of the soul, and one consequence of the approach of the infinite God, an approach that is his proximity. What a conclusion!

Does Levinas truly mean what he apparently says? Surely some kind of theodicy is present here, and certainly so if it is only an awakening to the horror of evil which makes possible the birth of our soul. Yet so far as I can see Levinas says very little here about evil, and can he really mean that it is only an awakening to the transcendence of evil that makes possible an openness to the proximity of the infinite God, or is this a misreading of his text? Can we only know the infinite God who is beyond being by knowing the transcendence of evil? But what is that transcendence? Levinas would certainly refuse any notion that the transcendence of evil is finally the transcendence of God, but how in this context can he avoid that conclusion? The preceding essay, "Humanism and Anarchy,"[4] perhaps sheds light on this question, for if all rationality amounts to the discovery of "the origin," and the intelligibility of the subject itself can consist only in a return to the origin—a movement which is the very "being of the ego," so that the reflexivity of the ego is nothing else than the fact of being the origin of the origin—this would appear to lead necessarily to our apocalyptic crisis, and a crisis for Levinas calling forth a new concept of passivity, a passivity referring to the "reverse" side of being, one on the hither side of consciousness and knowing, even referring to the antecedence of creation, the "metaphysical antecedence."

The pure ego, in its apparent absolute separation, which is the psyche, knows nothing prior to its sovereign freedom, and is obliged to be its own source. For the freedom of the ego is "beginning itself," yet this is a freedom and a beginning called into question by the "other," wherein the ego is brought to accusation, despite its innocence, by violence, even by the violence of the "other," one which obsesses it, and does so by imputing to it a responsibility, a responsibility for which it has not taken any decision but which, closed up in itself, it cannot escape. Reduced to silence it gives a response beneath the logos, one prior to the amphibology of being and entities or the Heideggerian difference, for now the subject is indeclinable precisely as a *hostage* irreplaceable with others. Now we can speak of the "beyond the ultimate" or the "pre-original" without it becoming ultimate or originary. This is a "hither side" or "pre-originary" which is a "subjectivity" prior to the ego, prior to its freedom and nonfreedom. For this is something that precedes the beginning and the principle, that is "an-archically *despite* being," for it reverses or precedes being. Hence it is a question of an unassumeable passivity, an obverse of being involving a "reverse" that cannot be turned over, for it is a responsibility prior to freedom.

Thus the true subject is a "preoriginary susceptibility," more ancient than "the origin," for the subject is a responsibility before being an intentionality. And this is true because the Good has taken possession of the subject before the subject had the possibility of making a choice. Just as there is no subjection more complete than this possession by the Good, this "election," this very subjecting character of an absolute possession is nullified by the goodness of the Good that commands, and the one obeying recovers his integrity on the hither side of subjection. And here (and only here?) ethics makes its entry into philosophical discourse, and does so in response to a responsibility overflowing freedom, with an obedience prior to the acceptance of orders. So it is that Levinas names the Good by starting with this "anarchical" situation of responsibility, in which one is dominated by the Good, and this is precisely to be excluded from the very possibility of choice. Yet this impossibility of choice is the effect of unexceptionable election by the Good, an election that is not an action, but nonviolence itself, releasing a passivity more passive than any passivity—a filial previous and pre-logical subjection—and it is this passivity that the Good *is!* Such passivity is the being, from *beyond being*, of the Good, one which language is right to circumscribe, betraying it by the word "non-being." For the "invisible" in the Bible is the idea of the Good beyond being, one obliging a responsibility for others, and it is the "trace" of a past that declines the present, the trace of an immemorial past. Is this that responsibility that is the true subject? A pure passivity preceding freedom, owing nothing to my freedom, for it is only a responsibility for the freedom of others? Nothing in this passivity of possession by the Good becomes a natural tendency, and it is an anarchical bond precluding the possibility of choice, for it is made without the subject having been a will. In this possession the difference between the possessor and the possessed disappears, yet it needs the temptation of the facility to make a break, the erotic attraction of irresponsibility, thus in the midst of submission to the Good the seduction of irresponsibility occurs. Then there occurs the very birth of the ego in the obeying will, and this temptation to separate oneself from the Good is "the very incarnation of the subject," or his "presence in being." Now evil is realized, and is so through an obedience that is free of servitude to the Good, an obedience to an *other*, and it is because of such an obedience to an "*other* that remains other" that the subject is "carnal" and "becomes a being." Here begins the axiological bipolarity, one in which Evil claims to be the equal of the Good, a Luciferian lie that is the very egoism of the ego, which posits itself as its own origin. Thus the being that perseveres in being is egoism or "Evil," a birth of hierarchy that is the birth of evil, a perpetuity that would claim to be eternity, one even giving us a

Hegelian "bad infinity,"[5] and surely giving us what Heidegger would unveil as the history of Being. Would it be possible to imagine a greater assault upon our history, and most clearly so upon our philosophical history, but one that is only manifest by way of our apocalyptic crisis?

Frankly, I am intrigued by these essays, and wonder if it is the very question of evil that impels Levinas's most radical thinking, a radical thinking driving him to that "pre-origin" that is beyond being, beyond consciousness, beyond knowledge. Is this that Eden that we have lost? An Eden beyond evil and therefore beyond being, and if here the advent of evil is the advent of everything we have known as being, and everything we have actually known as the subject, is it ethics itself that most deeply deconstructs our history, and deconstructs it by calling forth a responsibility that is absolutely prior to being, a responsibility that is quite simply our election by "the Good"? Is there any trace here of the *felix culpa*, one that would appear to be present in understanding an awakening to evil as the birth of the soul, or is that birth the birth of evil itself? And is it evil that is most ubiquitous in our world, or in "being" itself and the pure subject, a pure subject who is the very reversal of the true subject, and a "being" which is the very reversal of the Good? Is this a Gnosticism going beyond Gnosticism itself, or a true conceptual opening to the biblical God, that biblical God who is the one source of responsibility, and a responsibility wherein our subject or "I" can only truly be subject by being subjected to the Good?

Thomas J. J. Altizer

Letter 7

To Ray L. Hart* (April 13, 1998)

Dear Ray,

I have just completed the outline of your magnum opus, *God Being Nothing*,[1] and I hasten to respond, although I must immediately confess that I am not confident to whom I am responding, for this author is simultaneously overly present and overly absent, all of the Hartian signatures and nuances are here but now they are inseparable from either a deep absence or an all too subtle elusiveness. Or is the true Hart finally and fully speaking? As you confess in a footnote, the earlier Hart could not or would not think the Nothing or nothingness, does the later Hart finally intend to think deeply only *the* Nothing? Or that abysmally groundless pure indeterminacy that is the Godhead? Yet let me confess at once that I deeply admire such speculative theology, and all the more so insofar as it is so deeply absent today, perhaps being now alive only in your own mind and spirit.

As you know all too well, I can only genuinely respond to that which truly resonates with my own thinking and imagination, so I shall inevitably do that here. But first let me attempt to identify your source or origin that is so crucial to this work. Again you remain deeply grounded in a Neoplatonic tradition, one deeply alien to me, but clearly you are thinking both with and against this tradition, and perhaps more deeply against than with it. Your occasional wrestling with Hegel is illuminating, since you are deeply anti-Hegelian but nevertheless cannot escape a Hegelian language, or not wholly so. But then you seem to be distant from every modern philosopher, even including Heidegger as you point out, and yet this is in no sense a "postmodern" thinking, and cannot be so because it is a deeply speculative thinking. While you acknowledge your indebtedness to Eckhart and Boehme,[2] I wonder if you do not have a deeply inverse relationship to Plotinus, who could maintain that Matter must be described as Indefinite of itself, by its natural opposition to Reason-Principle (*Enneads* 2.4), a matter that for Plotinus is an utter destitution and an unredeemed evil. Here, matter is invisible and imperceptible, and insofar as Evil exists the root of Evil is in Matter. An inverse image of your Godhead? And surely your trinitarianism is not unrelated to that of Plotinus, here you seem to be far closer to Plotinus than to Augustine. Or to pass to the other extreme, I wonder about the relation of the late Hart to the early Neville,[3] you will remember that in *God the Creator*[4] (p. 114), he asserts that the divine creative power only comes to be in the actual creation, and the coming-to-be of the "determinations" is a novelty on the ontological scene, they come out of nothing, and are created through immediate power, which means that creation is eternal or not in time. Now, of course, you understand this as temporal rather than eternal creation, but nevertheless both of you seem to be equally affected by "determinations," and seem to share a quest for a Godhead without determinations. Notice that Neville's God is so transcendent as to be indeterminate with respect to the world, but nevertheless so immanent that there is nothing whose whole being is not the immanence of God. Nevertheless, I continue to be baffled by your origin, clearly Plato is primary, and a purely speculative Neoplatonism beyond that, but this is one that would appear to be uniquely your own.

I also am very much troubled by your relation to the Bible, a Bible that for you is only a "day-book," relating only to temporal creation, and presumably closed to eternal creation or the Godhead. This would appear to have genuine Gnostic overtones, or perhaps pagan Neoplatonic ones, and perhaps above all so in your insistence that the central "character" of the Bible is not there, being absent or off-stage. Now that is an extraordinarily odd reading of the Bible, as any biblical concordance would reveal, unless you are insisting that all of the biblical references to "God" are not references to God at all. Surely you intend to go wholly beyond the Bible, and is this an attempt to seek that Godhead that is truly beyond God, and thus necessarily beyond the Bible? But just who is that God who so continually speaks in the Bible? A pure cipher or surd? Do we best read the Bible apart from God? Here you remind me of contemporary biblical scholarship and now I can see why you so liked that PBS program "From Jesus to Christ,"[5] that wholly eliminated any kind of theological language, yet I would have thought that this kind of reductionism would be alien to you.

But let me at last pass to more fundamental issues. Immediately you declare that you intend to think the interior life of the divine God, and it would be difficult to imagine a more radical or audacious goal. Then we find an initial formulation of your primary theses, that God creates from the nothing internal to "godself," and that the eternal self-generation of God the determinate Creator is from the abysmal indeterminateness of the Godhead. So, too, your goal is to create a mono-"theistic" trinitarianism in "onto/meontogenesis." At once we are confronted with Hartian neologisms, which seem to be essential to your writing, and while you might affirm that the originality of your thinking demands this, I trust you are fully aware of its offense. You also quickly acknowledge your attraction to heterodox trinitarianism, indeed, you are committed to it, but then you cannot so lightly cast Hegel aside, who is surely our greatest purely heterodox trinitarian. Deeply unlike Hegel, however, and this is revealing, you affirm that trinitarian theology ventures to speak principally of the Creator, here you are orthodox, but nevertheless deeply radical or heterodox in subordinating the Creator to the Godhead, hence your deep ambivalence about Gnosticism.

Then we discover the truly startling claim that the highest aspiration of theology is to write the autobiography of God, one that despite Jack Miles[6] you see as being absent from the Bible, and now we discover that there are two lives of God: (1) the life of God exhausted by internal relations (immanent Trinity), God the determinetness of Godhead, who in that respect is external to the Godhead; and (2) the life of God the Creator who is God the Redeemer (economic Trinity). Here, as elsewhere, you say little about God the Redeemer, and all too significantly you say even less about the relation of God the Redeemer to the Godhead. Frankly, what I think you must most guard against, and you are deeply attracted to it, is an exoteric understanding of God as opposed to an esoteric understanding of the Godhead. Now since you most seek a theological ontology, one unfolding "meontology," or the pure and groundless indeterminacy of the Godhead, this threatens to overwhelm everything else in your thinking, and certainly to overwhelm anything we can understand as "God." Yet you also seek a theology of the interior life of God, and to do so by thinking systematically, thus determinately: "the determination process on which all determinate thinking, thus saying, depends." Apparently you most seek this by way of a fundamental distinction between *creatio ab origine*[7] and *creatio continua*,[8] and it is in the former that your deepest interest lies, so that your unique way is one of a backward determinate speculation, most deeply seeking that eternal creation that is the self-creation of God the Creator from indeterminate Godhead. Ironically, you ascribe this way to the Christian tradition, whereas most would surely see it as a radically heretical or gnostic tradition, and it surely cannot be found in any of the scholastic theologians, unless one were to identify Eckhart as such.

It is too bad that John Milbank[9] and his school have cornered the tag of "radical orthodoxy,"[10] for you seem to be seeking such, and not a Catholic variety like Milbank's, but rather an esoteric Protestant one, perhaps seeking a contemporary *Freundengottes*,[11] a community even now reflecting the "Emergence-y" of the divine God. For I think that you are most deeply interested in the genesis of God, which happens to be true of myself, and I often wonder how much you have read of my book[12] on the subject (surely not its chapters on Hegel which would have spared you the traditional interpretations of Hegel that you offer here). And you are not unlike Hegel in wanting to write about everything, which I think distracts the reader from your deeper subject, but I can see that this is systematically necessary. Now if what is crucial for theology is a theological meontology of the coming forth of the divine God, with the emergence of God from Godhead, you undertake to resolve this by explicating the *determination process* itself that antecedes any determinateness whatsoever. And this process is nothing other than eternal creation or the self-creation of God! Moreover, this is a trinitarian process, and one always and eternally trinitarian. Yet it most deeply arises out of the abysmally indeterminate *turba*[13] of Godhead, a groundless indeterminate *turbic energeia*[14] of the divine living/dying, or the formless energy in the divine depths (here fully paralleling Hegel's *Trieb*[15] who also derived this category from Boehme). You also join Hegel in thinking that this determination process is one upon which all ontological and meontological determinateness and determinate thinking depend, and for both of you this process is inseparable from an actual as opposed to an abstract nothingness. And both of you think that the Godhead is inaccessible to metaphysics, indeed, for Hegel the very thinking of this process brings metaphysics to an end, and Hegel thinks this through systematically in the *Science of Logic*. Yet for you there is no access to the divine abysmal indeterminacy save through the determinateness of God the Creator-Redeemer, whereas for Hegel it can occur only through the Passion or the Crucifixion of God. Once again as you are in the center of your subject, you are drawn to its peripheries, and tend to lose the real subject here. And in believing that Hegel's Spirit did not know its own indeterminate depths, depths insusceptible of self-mediation, are you affirming that such depths cannot be thought, or can be thought only apophatically? What then happens to your own project?

For surely yours is only all too partially an apophatic thinking, and certainly not in its attempt to think the *abyss* of Godhead, for you intend to call forth this depth unfathomable, and even to understand it as the "monanarchic groundlessness of trinitarian God." All too significantly it is at this very point that you claim the absence of God from the Bible, yet you here understand the default of language as the self-sacrifice of God that refers to God iconoclastically, leading one to wonder if you are here primarily speaking of biblical language. Why is the Crucifixion not an actual epiphany or an actual realization of the *abyss* of Godhead? Seemingly the Crucifixion is wholly absent from this thinking, which is certainly a profound difference from Hegelian thinking, but one might also think that Christ is absent here as well, and certainly Jesus Christ as opposed to a Neoplatonic Logos. Who or what is this triune Creator-Redeemer of whom you speak, one absolutely unique and without parallel in any determinateness of the world, and yet one wholly immanent, and the principal-agent of eternal and temporal creation out of principal groundlessness? How the concentration of both monanarchic indeterminacy and pluriarchic determinacy, and how the very Godhead of God? Is the Godhead of God now known as some kind of determinacy? Is this consistent with what has gone before? I can't think that your mind is really in these sentences, even if a Methodist heart is present, for you then declare that Godhead is the site or abode of the genesis of God, of the coming forth of divinely determinate God from indeterminate abyss, and is the locus of the "default" of determinate God. Is this the default of the triune Creator-Redeemer?

Now we proceed to your thinking about the *nihil*, one seemingly absent from philosophy until Schelling and Hegel, and one calling forth a Great Divide. On the one side Being, and on the other side the "kenary" depths of privation, and religions themselves divide on this momentous question. And our existing theological language at this absolutely crucial point says little about God the Creator and nothing about God the Redeemer, and cannot do so if only because it cannot know an actual or a real nothingness. Here, we are impelled to ask of the relation of Being and Nothing to God, but this is certainly not the perennial question in our philosophical theology, but only a uniquely modern one, for only modernity has thought the Nothing, unless this is true of precursors of modernity such as Eckhart, Lurianic Kabbalism,[16] and Boehme. You have a habit of identifying your own questions as perennial ones but this is simply not true historically as you well know. What traditional thinking cannot and did not think is precisely *nihilation*, but that is the very center of Hegelian thinking, and one to which you are unconsciously deeply allied. Yet do you fully accept such thinking? There is a baffling clause in which you seemingly speak of Being and Nothing as being determinate in the eternal self-creation of the divine living/dying, in the spiriting of the indeterminacy of the "Wisdom" of the Godhead. Then recognizing the truly new thinking of the Ground in the nineteenth century, you assert that it was occasioned by the project of thinking *Ungrund* meontologically, the very groundlessness of determinate being-cum determinate nothing. And this radically new project you accept in your own terms by attempting to think Godhead/God by thinking the determination-process, which you now identify as a metaxic "crossing" from indeterminacy (*Ungrund*) to nondeterminacy or *Abgrund* to determinacy (*Grund*) and back in the eternal living/dying that comprises the divine immanence.

Now we do move to the very passion of God, or to the crucifixion of God, but this is now understood as the withdrawal of God in the "crossing" back through to Godhead, the "self-refusal" of God for recuperative Godhead. You speak of this as the very withholding within "God godself," a staying-away of God from God the determinate Creator, and this is both an advent and a recusal of God that is unfinished. Surely this is a tomb that is empty, indeed, but is this an emptiness necessitated by a redemption of God wherein a determinate God, and one precisely therein in "default," is redeemed by returning to indeterminate Godhead, so that the abyss of Crucifixion is a renewal or *anamnesis* of the abyss of indeterminate or primordial Godhead? Notice the deep emphasis here upon "return," for it is an ultimately backward movement that is truly redemptive, and only a "backward" speculative thinking can recover such redemption. Of course, this is not recovered immediately or at once or even in a single act, for the inexhaustible indeterminacy of the divine Wisdom never achieves decisive determination, but this is nevertheless the very "economic" determinateness of God the Creator, a Godhead that has self-restricted its indeterminateness, thus the becoming and unbecoming of God by way of the becoming and unbecoming of the created order. Of course, this is a *monarchic* Creator, and this monarchy cannot be gone back on or crossed back through! Hence no real crucifixion of God the Creator, but nevertheless a withdrawal of the Creator, and a withdrawal back to indeterminacy that alone can resuscitate the Creator!

Ray, are you a reborn Valentinus?[17] Are you attempting to write a new Gospel of Truth?[18] This often sounds that way but your language is very much unlike that gospel if only because its every word was directed to the realization of that perfection that is absolute deification and your language most fundamentally and most deeply calls forth an abysmal nothingness that would seem to be the very opposite of a Gnostic deification. Perhaps this is occasioned by that far deeper abyss that has become incarnate in our world, but just as at least some Valentinians could know the crucifixion as apocalypse itself, you appear to employ the language of crucifixion in your conclusion, and do so by speaking of the "defaulting recusal" of God the eternally determinate Creator as a "crossing" back through for a replenishment of the indeterminacy of Godhead. Both you and the Gnostics seek an absolutely indeterminate Godhead, one that is itself the sole source of a deeper or ultimate redemption, and one that is the very opposite of what both Plotinus and the Gnostics know as Matter. Yet you can find a faint trace in the temporal creation of the eternally groundless ground of the divine metaxy, the "between" of the divine Advent and the Divine Recusal, and if here you employ the language of Voegelin[19] ("metaxy"), he only reached it by an ultimate conflict with Gnosticism, a conflict that you are clearly continuing.

Has such a conflict been with you from the beginning? Is that what your wrestling with Plato finally amounts to, or your virtual obsession with a seemingly Neoplatonic speculative theology, and your deep attraction to a truly heterodox trinitarianism? Perhaps the latter is most revealing, and it may well be that a truly new speculative theology can only be such a trinitarianism, one that Hegel certainly created, and one that Leahy has created in our own time. Perhaps I remain a simple Christian at this point, one obsessed with Jesus and the Cross, and you do have an indifference to the Bible that truly amazes me, and while the biblical theologian has commonly thought that this is inevitable for the philosophical theologian, you seem to have no bad conscience at all here, and can proceed as though the Bible has no relevance at all to deeper theological questions. Now that is not true for the poetry and art that you revere, and you know it, and I would add that it is not true for our greatest modern speculative theologian, Hegel. Indeed, it is not even true for Nietzsche, and I wonder if your reluctance to speak of the death of God is a reluctance to speak of the Bible itself. Perhaps this is essential for your work, and perhaps you are giving a deep biblical witness, you surely do indirectly speak of the crucifixion in your closing pages, and perhaps it is only here that you reveal your understanding of redemption as a return to the abyss of an absolutely indeterminate Godhead. And I suspect that you most deeply understand the Nihil as that abyss, one far deeper than that Nothing that is the correlate of Being, and far deeper than anything that we can know or name as God. This is a Nihil that is far deeper here than what you name as the Creator-Redeemer God, and if the default of that God can only be redeemed by a return to the abyss of the Godhead, is this that "recusal" that is not only the paradigm of redemption but the only possible source of true life or joy? And is this the absolutely crucial point at which we must effect an absolute *epoche* of the Bible because it lies beyond any possible language?

With deep admiration and gratitude,
Tom

Letter 8

To Thomas A. Carlson* (December 30, 1998)

Dear Tom Carlson,

I am deeply grateful for your gift of *Indiscretion*,[1] a title belied by the book itself, this is a marvelous book, a major theological achievement of our time, a time I had feared was closed to such work. So this is a deep gift, indeed, and even one seemingly "impossible," at least from my jaded perspective, and while I am proud to have been included within your horizon, I fear that I cannot now enter this promised land. The real problems of this book lie with your expositions of Hegel and Heidegger, for I fear that I cannot take either Derrida or Marion[2] with deep seriousness, and I suspect that you have been far too dependent upon Marion and von Balthasar[3] for your understanding of Dionysius,[4] whom I can regard only as a profoundly Eastern and Neoplatonic if not Gnostic thinker, but then such thinking is truly alien to me. But at least Marion and von Balthasar do not divorce Dionysius from the Church as you seemingly do, and I rejoice that Marion can only know Dionysius through the absolute authority of the "bishop," an authority that you wholly dissolve, but this dissolution is finally a dissolution of history itself, which is perhaps necessary in a genuinely postmodern theology.

Now yours is unquestionably a deeply serious theological project, and one that even parallels my own, although you are far more responsible than I am in your expositions, which are certainly scholarly as mine are not, even if I must deeply resist your understanding of Hegel and Heidegger, and most resist it for deeply theological reasons. But first let me say how much I admire your expositions, you have brought real clarity to an extraordinarily difficult thinking, your mastery of this thinking is truly admirable, and you ingeniously conjoin the original text with many of its major commentaries, and in such a way as to make these texts meaningful now. Indeed, the power with which you call forth Hegel and Heidegger almost has a comic resolution in your closing sections, for even if we bracket Dionysius in this context, Marion and Derrida in this perspective stand forth as truly minor thinkers, or, at least, the later Derrida does so, and this does result in an anti-climatic ending, which is perhaps saved by your own project. Yet what is most important here is your major conjunction of Hegel and Heidegger, one that I see as a seminal philosophical project of our time, and while I understand Heidegger as being deeply Hegelian as you do not, and I mean Hegelian in the sense that Marx and Kierkegaard are Hegelian, I also see Heidegger as being far more theological than you are willing to accept, and finally perhaps just as theological as Hegel himself.

Now we each share a common center in the death of God, and while I think that you are unwilling fully to enter this center theologically, you do so philosophically, and it is this very disjunction that not only most separates us, but perhaps makes possible your project. I deeply resonate with your centering upon "death" in the death of God, and your acceptance of "death" as our ultimate category, and not only an "existential" but just as deeply an ontological category, one that I believe is closed to all forms of ancient Western thinking, but that ever increasingly overwhelms all truly modern and even postmodern thinking. Yet it is just here that I think that you most deeply miss or annul both Hegel and Heidegger, just as I think that the gulf that you here establish between them is an illusion, and most clearly an illusion theologically, for it is just at this point that each has shattered our deepest theological tradition. As I think upon this matter, I think that the crucial problem here is that of "finitude" itself, and I suggest that first this problem must be understood historically, thereby banishing all ahistorical understandings of finitude. I think that what we understand as finitude has no correlate in the ancient world, and this because the very idea of "infinity" as we know it was only born at the time of the genesis of Christianity, there is nothing in Greek philosophical thinking approximating what later became established as infinity, and the very words *to apeiron*[5] in the Greek philosophical lexicon have wholly pejorative connotations, except for its one positive occurrence in Anaximander.[6] And when the thinking of infinity was truly born, it almost immediately was known as an absolute infinity, and an infinity absolutely other than finitude, thereby giving finitude itself an absolutely new identity. Such thinking truly ruled for well over a thousand years, only truly being challenged by Islamic philosophical thinkers, but this challenge gradually penetrated the West, thence creating the Thomistic synthesis, and here Marion and others are right about Aquinas. But it is also true that such a synthesis itself evolved historically, even if being virtually reversed by radical nominalism, and just as it is possible to see a genuine continuity between medieval and early modern thinking, despite the revolutionary ground of that new thinking, and one itself probably made possible by nominalism, it is also possible to see a profound transformation here, and a radical transformation of a millennium and a half of thinking, as most purely present in Spinoza's revolutionary identification of infinity and finitude. While Heidegger deeply ignores Spinoza, Hegel was profoundly affected by Spinoza, and above all so at this point, and it is here that the negative thinking of Spinoza and the negative thinking of Hegel deeply coincide. Now this is a history that I think that you deeply miss, and this is a crucial reason why I think that you misunderstand Hegel's understanding of finitude, which you approach much too abstractly, but even here you miss Hegel's profound exposition of infinity and finitude in the *Science of Logic*, and thus miss his dialectical identification of finitude and infinity.

For the Hegelian negation of finitude is truly an *aufgehoben*,[7] which is to say that finitude is preserved in that negation, and not only preserved but enriched. Now a deep theological model of this is the Crucifixion itself, for the Crucified God in being the negation of the "infinite," or that infinite that is simply the opposite of the finite, is at bottom the realization or actualization of the depths of finitude and infinitude at once, so that thereby a truly new finitude and infinitude are finally called forth, but now and for the first time one in which there is no true difference or otherness between finitude and infinity. Again and again you speak of an Hegelian negation of finitude as a non-dialectical or non-Hegelian negation, thereby falling back to an "idealistic" understanding of Hegel, which I had thought could never actually be possible again. Now it is true that such an understanding does appear at times in his lectures on the philosophy of religion, but let us always remember that these are not an Hegelian "text," we can have no confidence that they represent the real thinking of Hegel, and I even believe that the late Hegel is a betrayal of the mature or real Hegel, and above all so in his social and political philosophy.

I also believe that Heidegger's understanding of finitude must be understood in such an historical context, for just as Heidegger knows that his understanding of finitude is only made possible by the ending of metaphysics, that ending is itself the culmination of two and a half millennia of thinking in the West, and as few Heideggerians apparently understand, it underwent deep transformations during that long period. Now I do admire the deep clarity that you bring to Heidegger's understanding of finitude and death, although I think that this must inevitably be a betrayal of Heidegger's own thinking, which profoundly resists or transcends all such clarity. And I fear that you impute a literal meaning of death in Heideggerian thinking, as though death is quite literally an ending and only an ending, and therefore nothing that could be actual for us. Here, you profoundly miss not only the deep grounding of Heidegger in poetic language, and Heidegger was profoundly affected by Rilke's *Duino Elegies*,[8] which ecstatically call forth an ultimate identification of life and death, but also Heidegger's deep ground in the Crucified God, for even if this has previously been largely hidden, it is now being called forth. I just happened to read an excellent study of this by Orrin Summerell, in a volume he edited for the University of Virginia Press entitled *The Otherness of God*.[9] I even think that the Crucified God deeply even if hiddenly underlies Heidegger's exposition of "being-toward-death" in *Being and Time*, note that *Dasein* reaches its "wholeness" in death, and this is a gain that is an absolute loss of "being-in-the-world," for never again can *Dasein* be experienced as "a being." Indeed, "ending" as dying constitutes the totality of *Dasein*, and while essentially there can be no "representation" of this ending, that certainly does not mean that it cannot be actual for us, an actuality that can be present in an existential understanding of death. This, indeed, calls forth Heidegger's understanding of fallenness and Angst, and so, too, his understanding of an actual nothingness, an understanding that Hegel conceptually created, and at this point Heidegger is in deep continuity with Hegel.

You ignore this deep dimension of Hegel, and therefore miss a crucial link between Hegel and Heidegger, but I believe that it is overwhelmingly important that Hegel and Heidegger are our only philosophers of an actual or real nothingness, unless this is true of Schelling and Nietzsche. Of course, you also miss Heidegger as a manifestly theological thinker by so ignoring the late Heidegger. Is this essential to your project? Why? Can it be because this Heidegger as a manifestly theological thinker is at such an infinite distance from apophatic theology?

I must confess that in this perspective Heidegger is for me more deeply Christian than Dionysius, for despite your analysis I can only know this theology as one of the Christ of Glory, a Christ who is even more deeply absent in Heidegger than in Hegel, yet in our world Dionysius does have the deep advantage of being such a profoundly alien thinker to us, I sense that your attraction to Dionysius is not unlike mine to Buddhism, and perhaps such a truly dialectical conjunction will play a comparable role in your theological development. Certainly I hope that this will continue, for you have made a truly fundamental and exciting beginning, and I am in awe of this achievement if only because it has occurred in a world that is so alien to theological thinking, that I did not have to contend with, thus I cannot really understand it, but I do deeply admire it.

Thomas J. J. Altizer

Letter 9

To D. G. Leahy (April 4, 2000)

Dear David,

This is a response to your recent public lecture,[1] a truly powerful one, and given its author amazingly clear, although so much is here that it is very difficult to apprehend how much of it one has grasped. First, let me commend you on its dramatic form and power; is the original epic poet[2] here resurrected or renewed, or is this the advent of a new voice, and precisely as voice one with a universal impact? Of course, this is the first time that I have heard you present original thinking in an oral mode, but as a witness to it I can testify to its immediate impact, and one I suspect even affecting those who were most closed to it. Lissa and I deeply regret not having brought a voice recorder to record this event, and I do think that it was thoughtless for the sponsors of the event to have neglected this, but perhaps few are aware of the power of your voice, one that I have long believed gives your thinking a universality that otherwise is absent.

This lecture has a remarkable and I believe unique beginning for you, it is disarmingly clear and decisive, and it does truly open the way for what follows. Yet it also evokes the genuine mystery of the "I" as subject, one that cannot appear in the actual now, being beyond both presence and absence, and you deftly draw this forth as the groundlessness for the very notion of selfhood. Yet it is precisely therein that you would apprehend the I as the transparent foundation of freedom, even if you baffle your hearer by so immediately moving to an affirmation of the Mother of God's conception as the apocalypse in embryo, and thereby move to an apocalyptic discourse. Of course, this whole lecture is apocalyptic discourse, but never before has yours begun with calling forth an immediacy that all can understand, hence evoking the apocalyptic as that which is most immediately at hand. And by at once affirming that the apocalypse is the "end" of the end of the world, and thereby that new universe that is the truly first "now" of the world, for this is the beginning of the beginning of the "now" of the world, one present in this absolutely new way of thinking, and one essentially uncovered in the apocalyptic

"I," and an apocalyptic "I" because it is the end of the human "I," and you do give us a new and decisive way into that thinking itself.

There follows what I take it as a new way of understanding our Christian history, one divided into three millenia, and the millenia themselves here apprehended by way of the names of Augustine, Aquinas, and Leahy (or if you prefer, "the thinking now occurring for the first time"), moreover, there is a clear progression establishing an evolutionary understanding of the movement of the millenia, a movement that is a necessary and essential movement, and one culminating in what is clearly a final age, or that which has traditionally been known as the third Age of the Spirit. I also am intrigued by your apocalyptic identification of the three ages or millenia, the first wherein the mind is materially but informally apocalyptic, and the world as inessential substance, the second, where the mind is formally apocalyptic, as transcendental although natural reason, and the world as essential substance, and the third, the purely apocalyptic, with complete objectivity, the absolute transcendence of knowledge, and the identity of conception and perception, of acting and thinking, of imagining and accomplishing.

My only real problem here is with your understanding of Hegel, and specifically your understanding of the identity of becoming in the Logic, for you ignore Hegel's deep centering here upon the advent of becoming as a consequence of the negation or self-negation of an original identity of being and nothing, for it is the vanishing of that pure being and pure nothing that is the advent of becoming. You commonly ignore the crucial role of the "nothing" in Hegel's thinking, and as you know I believe that the "nothing" did not enter pure thinking until the advent of German Idealism, except for its prior and ultimate role in Buddhist thinking. So, too, I think that it is just the "nothing" that is most ambivalent and elusive in your thinking, but more of that anon. And while I have no real problem with it, I think that your discussion of anxiety here would have been illuminated by calling forth the "nothing," one that is decisive for both Kierkegaard and Heidegger, but this is not truly post-Hegelian thinking, as witness the crucial role of the "nothing" itself. I sense that it is really only Levinas whom you can discuss within this horizon, and I grant that he is post-Hegelian as Heidegger and Derrida are not, but as you so decisively show this is possible only by seeking a subjectivity prior to consciousness itself. Incidentally, I like very much your identification of the American death of God theology as turning the Godhead inside out, that states my project more clearly and decisively than I have been able to do.

Yes, I fully agree with your proposition that is the Idea of God that has truly precluded the apocalyptic "I" of God, although I cannot agree that it is only the modern Idea of God of that this is true, since as you know I understand the whole world of Christian scholasticism, including both Augustine and Aquinas, as effecting that occlusion. And this is occluded by the absence of an absolutely dialectical movement in all scholasticism, but how can you say that is only the essentially new form of thinking itself actually beginning that exhibits the dialectical element in the principle of absolute idealism, and doing so precisely by actually existing now for the first time as the beginning and end of the principle of modernity? How is the end of the principle of modernity the absolute clarification of "the beginning"? Does this not essentially occur in absolute idealism itself? Or is this impossible for that idealism because of its centering upon an absolute self-consciousness? And is it the negation or self-negation of that self-consciousness that releases the consciousness of the beginning before the beginning of consciousness? Now you do make clear how close you are to Levinas, even seemingly claiming that your thinking achieves that which Levinas could only attempt, a consciousness otherwise than knowledge but not otherwise than being, a consciousness otherwise than the beginning of being but not other than "being and nothing," a consciousness of the beginning not becoming, and here a ground of your misreading of Hegel's notion of becoming becomes clear. And to identify this beginning as the beginning of consciousness as complete objectivity? Is this Hegel turned inside out?

How marvelous that this deeply abstract language is followed by your all too concrete discussion of the ontological donut. I think that never before has your language been so concrete, and it is even more so in its oral expression. And how exciting it is that this language centers upon "the nothingness of nothing," even calling forth that nothingness as "the being of everything beginning," as fully made manifest in the essentially new form of thinking that *is* beginning. And not only is actually beginning now, but that itself is absolute beginning, and is absolute beginning because here the nothingness of nothing is the being of everything beginning. Moreover, this is truly made concrete by calling forth the Apocalyptic God, one revealing or making manifest that the nothing in the middle—the hole in the donut—is freedom, and even human freedom. This is the freedom that has been consumed by God, but not thereby wasted, and this consumption of "nothing" is the very undoing of the Idea of God. All of this, of course, is most deeply directed against Hegel, and you can even claim that exhibiting the dialectical element in this nothing wasted is the thinking now occurring for the first time. But I fail to see how this does not truly occur in Hegel, and most manifestly so in his radically new understanding of nothingness, for the first time understanding nothingness as an actual nothingness rather than a privative nothingness, and an actual nothingness already realized or actualized in the original self-negation of Absolute Spirit or Godhead itself. It is precisely that nothingness that makes possible an absolute negativity, a nothingness apart from which Spirit would be lifeless and alone, as Spirit finally is in every non-Christian apprehension of the Godhead. You also make a strange slip when you speak of God dying after Hegel, why ignore the centrality of the death of God in Hegel's deepest and purest thinking? Yes, this theism is certainly atheism, for all modern theism is finally atheism, and I would here include scholastic theism as well. And when will you examine Fabro's glorious book *God in Exile* on modern philosophical atheism?

I sense that it is the very category of nothingness that is most important and decisive for this lecture, and most revealing of its apocalyptic ground, one only fully drawn forth when you speak of the Apocalyptic God consuming everything and wasting nothing. Here, I am most intrigued by your language about that something that is left to its nothingness when everything else including "nothing" is not left to its nothingness. Now you do speak clearly of what in logical terms you call the "unum," but now you can speak of it as the "I" that is the beginning and end of the beginning, and even speak of it as the beginning and end of the donut that is this body and your body! Are you for the first time speaking concretely of the body? This is the "I" that survives when everything including nothing is consumed by the Apocalyptic God, and if you again employ Pierce's[3] understanding of an everything including nothing, you now clearly set forth a something excluding nothing, and now this is the "absolutely objective being of the I otherwise than presence." Yet once again I see a crucial distortion of Hegel, and this in your affirmation that the Hegelian God is the truth of being and the I but not the truth of the nothing, and how fascinating that this is the context in which you now affirm that the God of the essentially apocalyptic thinking now actually beginning is the truth of being and nothing but not the truth of the "I." In your footnote, you even affirm that this is the opposite of Hegel, and I think that this is true in this language about the I, and perhaps this is just the point at which your distance from Hegel can most clearly be apprehended. Of course, I would say that this can most easily be understood by noting that deep dissolution or self-dissolution of the "I" of self-consciousness that has historically occurred between Hegel and Leahy. Now we learn that the true I is the essence of the irreducibility of the body to nothingness, the I that is the core of the emptiness at the center that is no center of the body or the world, the core of the body's and the world's not having a center, and even the "person" or I that discloses itself to itself immediately in the unreflected objectivity of the body and the world. This truly is non-Hegelian, truly does evoke a third millennium, even if it is a third millennium that truly threatens to pass into the absolutely primordial. As you know, I identify the absolutely primordial as our deepest threat and danger, one that can be identified as the truly pagan, and one that has erupted more fully in our century than in any previous century of the Christian era.

Now if the something excluding nothing from everything is not consumed by the Apocalyptic God, and if this is the "I" that is the perfectly transparent surface of the body, or that finite spirit that is the actual boundary of the body, the I surfacing the body, how is this I the distinction of the not wasted being and the not wasted nothing? Just what does "nothing" mean here? Is it this "nothing" that marks the truth of this I's freedom as the divine freedom, one given in the Creation, but only truly manifest in that thinking now occurring for the first time? Here, you can even affirm that the truth of this I is a "nothingness" for the first time otherwise than owning, beyond the very notion of the self. For the I acting absolutely *is* the reception of the gift of being itself, the I that is the infinitely transparent surface of the body is absolutely the form of the absolute gift of being. But is this only actually made manifest by this nothingness itself? Note that this I is simultaneously invisible and the infinitely, transparently, surface of the body, an I wholly different from the me, and yet an I that is an apocalyptic I, and perhaps most clearly so as an invisibility that is the wholly transparent surface of the body. I am also intrigued that this is the point at which you introduce your fullest discussion here of Levinas, virtually apprehending his thinking as your true predecessor, and showing how his thinking is only truly resolved in your own. Now I can see that Levinas's understanding of the "preoriginal me" is truly important, and I can see that it is extraordinarily important for you to establish an absolute distinction between the apocalyptic I and every "preoriginal" identity, one most fully given us by Levinas himself. Perhaps this can most decisively be established by reversing Levinas' ethical thinking, and doing so by evoking an ethical spirit immersed in the "waters of ontology," and if this impels you once again to speak of the I of the apocalyptic consciousness as being essentially ontological, phenomenological, and transcendental, is that not clearly Hegelian language? Perhaps this becomes most concrete in your very language about essence, for even if essence is the exception to essence, how does this truly differ from an Hegelian understanding of essence wherein essence is the absolute negativity of being, and is so only insofar as being has negated itself, and negated itself insofar as it is infested with otherness, for it is the perishing of that otherness that is the realization of essence (*Logic* 2.1.1). Now it is true that if you could really call forth a "physical ethics" or a "bodily ethics," that this could be truly non-Hegelian, but I think that this is just what Nietzsche accomplished, and did so upon a deep even if hidden Hegelian ground. Certainly both Hegel and Nietzsche created an ethics of the order of actuality, that is the deepest offense evoked by each in their radically different ways, and so, too, did each create an ethics inherently disruptive of the world order hitherto existing, an ethics essentially transformative of the order of reason existing hitherto. But Levinas as a deeply primordial thinker, stands wholly apart from history, or can relate to it only as a history that has come to an end.

If it is this situation that occasions or makes possible the rupture of the totality, a rupture that Nietzsche so deeply understood, and if this rupture is caused by an explosion of the order of the world itself, an explosion that you identify as the "absolute explosion of exteriority," this is an explosion that is apocalypse itself, an apocalypse that is a completely new existence that we can all feel in our bones, a beginning that is the absolute judgment, one calling forth for the first time the transcendental dichotomy that is now the actuality of the ego itself, a transcendental ego experiencing for the first time the absolute reality of the end of the world, and also and perhaps even thereby experiencing for the first time an absolutely manifest God. Yet now and for the first time the "I" I now am, devastated of being and nothing—the ontological donut downed—is absolutely objective and completely free. This is that now actually existing person who is the foundation of a society that is the beginning of an essentially new world, or the beginning of apocalypse itself.

Your final paragraph, or your closing homily, is certainly a celebration of that "being and nothing" that now and for the first time constitute the complete matter of the body's consciousness. Now the infinitely transparent I is the surface identifying body and world absolutely, and now we are given a real society of categorically free I's, a society of wasted somethings that is not itself wasted, is something consumed by God after being and nothing have been consumed. Indeed, this is the threshold of the third millennium, that apocalyptic era that is heaven incarnate on earth, or that paradise that previous visionaries could only dimly perceive.

Never have I before heard such Christian preaching, or such apocalyptic preaching! Yet I fear that in my perversity I can only hear it as real insofar as you evoke the "nothing," and this despite the fact that this is a nothingness that here seems truly ambivalent and elusive, truly beyond language itself, and truly beyond our understanding. Perhaps that is what makes it most real to us, and if this is just the point at which this discourse is most offensive, is that not an offense essential to this discourse itself, and that very offense that most engages us, or most calls forth our genuine hearing? The hole in the donut! What a marvelous image of our apocalyptic "I" or of apocalypse itself!

Tom

Letter 10

To Cyril O'Regan*
(April 12, 2000)

Dear Cyril,

I am deeply grateful for your spirited response to my recent ventures, as once again I know that I am in the presence of a truly major theological mind, I only wish that I could attempt to respond to your recent work. Yes, as I grow older I am increasingly intrigued by the deep relationship between Gnosticism and apocalypticism, coming to believe that this is a decisive key not only to Christianity but to our actual history as well, just as I will never lose the impact of a Bultmannianism that demythologizes apocalypticism and Gnosticism simultaneously. I am also moved by your suggestion that my apocalyptic theology represents the therapy for a complex set of diseases, and no one is now unraveling these diseases so ably as you. Yes, apocalyptic or genuine apocalyptic is characterized by a revelation that is fundamentally interruptive, inaugural, and dialectical, and here it is opposed both to orthodoxy and to Gnosticism. I would like to think that my criteria are very clear although perhaps only you can see this. And I am fascinated that each of us is so drawn to a common body of discourse, although from deeply different points of view, and if you are an "orthodox" theological thinker, you are the only living one whom I can now genuinely and deeply respect.

Yes, I am seeking the meaning of a forgiveness truly transcending and liberated from all moral dimensions, and doing so through the category of transfiguration, which unfortunately continues to elude my thinking. And, yes, transfiguration is here a "meontological" power and precisely as such apocalyptic, although this word is alien to me if only because of its Neoplatonic overtones. No doubt Boehme is deeply important here, but I await a genuine initiation into Boehme, which hopefully will come from your work, and it is the early or middle Schelling who has affected me and not the late one, just as Berdyaev[1] much affected me in my earlier work. You are right that Ray Hart is working in this arena, centering upon an absolute nothingness, but none of his real work here has yet been published, and if you should become interested, I could send you a copy of this work in a provisional form. I also am not surprised that you, too, are drawn to the harrowing of Hell, which I think of as one of our most forbidden theological topics. But I hope that my use of this trope is first-order and not second-order, since I believe that this has been profoundly hidden from us, but I resist your position that the harrowing of Hell is found only in Blake and not in Dante and Milton. First, because I think that the *Inferno* is essentially and integrally related to the *Purgatorio* and the *Paradiso*,[2] and second because I think that Satan and the Son of God are essentially and integrally related to each other in *Paradise Lost*,[3] and that this relation is finally a truly dialectical one, for Satan and Hell are absolutely necessary to a Miltonic salvation, so that here damnation and salvation are essentially related to each other, just as they are implicitly in the *Commedia*.[4] And let us not forget the harrowing of Hell in *Finnegans Wake*![5] And, yes, von Balthasar's *Mysterium Paschale*[6] has much affected me, I only wish that I could have had a conversation with him.

Yes, orthodoxy is a Gnostic return, but an inverted one, and while it does truly negate Gnosticism, and becomes orthodoxy in doing so, its negation is more like a repression, and the repressed continually returns.

I must confess that despite my recent writing about theological orthodoxy I am uneasy in this engagement, largely because I cannot discover a contemporary orthodoxy to which I can respond, and once again I am counting upon you. The apocalyptic tradition, or its modern expression, as the privileged site of "Gnostic" return? That is a truly exciting theological venture and one that I presume can be realized only by you! But I also cannot resist asking if this is true in primitive Christianity as well, and here you yourself may well be under the impact of a truly theological demythologizing, and let us never forget that Bultmann knew this as occurring most deeply in the New Testament itself.

I rejoice that you so respond to my reading of Spinoza, but I did not read Deleuze's two books on Spinoza[7] until after I had reached my own interpretation—incidentally it is Deleuze's big volume on Spinoza[8] which is the only book of his I truly respect. And I do wonder that Spinoza has so disappeared from the theological world, and even from the philosophical world, it is a scandal that Heidegger ignored him (once again anti-Semitism?), for it was Spinoza who first deconstructed the onto-theological tradition, unless this already occurs in a radical nominalism. Another primal thinker who ignores Spinoza is D. G. Leahy. Have you yet entered his work, you must if only because it is so deeply Catholic and radical at once, and if you like I could send you a copy of a recent lecture of his, which I think is the best brief expression of his thinking.

Incidentally, since I last wrote you a publisher has accepted my theological memoir, and will publish it under the title of *Living the Death of God: A Theological Memoir*, the title being their choice, and while I can live with it, it gives me qualms because it seem so presumptuous. I shall send you a copy when it is published next year.

I trust that you are aware that I am now retired and living in the Pocono mountains in northeastern Pennsylvania, 95 miles from Manhattan. While I know nothing of your personal situation, if you or your family are near this area (this summer?), I would be delighted if you were to visit our mountains and my home, Leahy has a summer home about ten miles from mine.

Tom Altizer

Letter 11

To Gan Yang* (September 8, 2001)

Dear Gan Yang,

I have intended to write you for sometime but an occasion has now arisen in which I feel impelled to do so, and that is a very concrete realization of the religious-political crisis which is now so real and so little understood. It is true that I am at least temporarily blocked in my present writing, although I am engaged in serious projects. Yet a few recent encounters have somehow affected me, leading me to realize how extraordinarily difficult it is either to think about or to discuss our new political-religious world, which at least in America is a truly opaque though clearly very powerful world. Of course, I cannot speak for China, but I sense that a parallel may here exist between China and America, but one wholly unexplored, at least as far as I know. First let me say that my own way into this problem is primarily by way of "Satan," at this crucial point I am a unique theologian, but if you are interested in my response here you should be aware of this ground, which is perhaps most clearly presented in my confession entitled, "A Vision of Satan."[1] Did I send you that paper? Perhaps not, so I shall include it as an attachment to this letter.

For I am persuaded that the "religious right" is truly Satanic, not in the common sense of that word, but rather in a critical theological sense, one that identifies "Satan" with the absolute No-saying of God, an ultimate and absolute No that reverses all life and energy, and is the ultimate source of what Nietzsche identified as *ressentiment.* This occurs in a context that Lissa persuaded me that I share with you,[2] and that is an irresistible sense that we are now living in a final crisis, a truly eschatological or apocalyptic one, in which everything that we can actually affirm is truly vanishing, and a dark void is consuming us all. I believe that for the first time in history such a void is truly universal, and while it is virtually unnamable, it is known by all who are awake, while simultaneously being unspeakable and if only thereby invisible, and above all invisible for those who most shape our public consciousness and our common thinking and understanding. This is most concretely manifest in the political arena, at least in America, and above all in that truly new and universal conservatism which so dominates our "postmodern" world. Recently I have been thinking about the integral relationship between this conservatism and our new theological conservatism, one dominating the theological world, and not only the Christian world, but all religious worlds. Critical theological thinking is now apparently coming to an end, both in our churches and in our academic worlds, and I think that for the first time in history, no ultimate challenge of any kind is now occurring, or insofar as it occurs, it occurs only in fantasy, or in academic games. Has fantasy ever been so pervasive as it is today, or has critical thinking ever been so removed from actuality, or our imagination ever so closed to the actual possibility of transformation? I believe that there is now an overwhelming challenge to name our darkness, and thereby I am very much interested in your response to this situation.

Of course, to employ the name of Satan in such a venture is seemingly the purest fantasy, but if only through the name of Satan we can speak of an ultimate or absolute evil, and I am persuaded that if we cannot speak of that evil we cannot speak of our darkness, just as I am equally persuaded that if we cannot know the absolute negativity of the Godhead we cannot know the depths of our darkness, or the ultimacy of that dark void which is consuming us. This is the context in which I have become fascinated by the enormous power of the "religious right" today, at least in the West it has been the name of God which has been our most ultimate sanction and most ultimate authority, and I am coming to think that modern secularism has finally deepened even as it has truly disguised that name, so that in times of ultimate crisis it is truly reborn, but it can only be so reborn as an absolutely repressive power. When Blake named God as Satan, he was most clearly so naming the God of late Christendom, that God whom Hegel knows as abstract Spirit or the "Bad Infinite," but this truly is the "God" of the late modern world, and a "God" who is truly being resurrected or renewed in postmodernity, but now renewed in a universal world, and a universal world of universal emptiness or darkness. Why is it that virtually no one attacks the religious right, not even spokesmen for the major churches dare do this, for now a new fundamentalism is truly universal, and the Bush administration is treating this fundamentalism as its major source of political power, and not even liberal Democrats dare challenge the fundamentalists. Everyone knows the mindlessness of the fundamentalists, and their deep opposition to everything which we know as a higher culture or a higher learning, but what is perhaps most fascinating is their deep alliance with traditional or established political conservatism, as though it is through that conservatism that "God" can truly be renewed and reborn, and this is the "God" whom I would name as Satan.

Now this is not possible unless it is genuinely possible to know God as Satan, and I am increasingly persuaded that this is something which we all at least implicitly know, and if properly decoded we could understand a uniquely modern philosophy as enacting such an identification. In reading *Beiträge*,[3] and Heidegger's attacks therein upon the Christian God, I could sense that Heidegger finally knew this God as Satan, and while he would never employ such language, if only in this text which he himself would not publish, he is forthright and passionate in his assaults upon God. But do not such assaults inevitably occur, whether directly or indirectly, in all of our major modern philosophy, and throughout all of our truly major modern literature? And are not such assaults responses to a truly alien transcendence of God, one it is true known in ancient Gnosticism, but as opposed to every possible Gnosticism full modernity knows and enacts an alien "God" which incorporates or embodies the Godhead itself, and is it not our condition that we inevitably speak of such a "God" when we evoke the name of God itself? But this is no abstract problem; it truly is our "existential" situation, and one which is surely becoming universal today, as witness the virtually universal reactionary movement of those who actually evoke the name of "God" in our midst. Is it possible that religion itself can only be reactionary or repressive in our time, which I believe is a unique historical situation, and does this go far beyond everything which we can understand as religion, for it seemingly pervades our consciousness and society as a whole, as witness the absence today of every real challenge to what Hegel understands as the "Given," or every real challenge to our actual society and consciousness.

This is a crucial point at which I seek your response, for I have long been persuaded that everything that the West understands as "God" is absent from Chinese and Far Eastern tradition, just as I think that Confucianism is the first genuine secularization that occurred in the world, here once again being under the impact of Weber, and if Confucianism in one form or another is a genuine core of all Far Eastern culture and society, is "God" absent here as nowhere else in the world, and does this foreclose the possibility of reactionary politics in the Western sense? Indeed, my real question is what we know as reactionary politics in the West today, or our new conservatism itself, inseparable from "God," and is this most clear from a Far Eastern perspective? I suspect that it is extraordinarily difficult for a Chinese intellectual to understand religion and politics in America, although this is probably not true of you, but are America and China true contraries or opposites at this point, for if religion has been a driving force in America as it has been nowhere else in the modern Western world, is the very reverse of this true of China, and above all of contemporary China? I think that ethics is a revealing discipline here, and if China created the first truly secular ethics, did America create the first truly secular religious ethics, and one impossible apart from its religious ground, and is the contemporary crisis of ethics in America one deriving from this realization, and is this why Levinas is now apparently the most influential philosopher in America? Could Levinas mean anything at all in a Chinese context? And if Levinas's ultimate call is to an absolutely primordial "Infinite," is such an Infinite totally meaningless in a Far Eastern context, but truly real in a Western context that ultimately knows Godhead itself as primordial Godhead, and is incapable of knowing the absolutely primordial as anything other than Godhead itself? Note that the return of theology in recent French philosophy is the return of an absolutely primordial theology, one truly distant from our Western theological traditions, and one embodying an ultimate movement of eternal return, an eternal *return* which I sense is impossible in Far Eastern traditions, but an eternal *return* which is the ultimate ground of every reactionary movement, so that primordial thinking in a Western context is inevitably a reactionary thinking, as witness Heidegger himself. Is our world, or our Western or postmodern world, one that can only ultimately hear the call of the primordial itself, and could it be that we are now more deeply bound to "God" than ever before? Now even if this is almost wholly an unconscious bondage, is this the ultimate source of our contemporary emptiness, and the ultimate ground of our profoundly and pervasively conservative world?

For I would correlate our new emptiness and our new conservatism, an emptiness not only driving people to that conservatism, but one creating that conservatism itself, which I would see as being unique historically, and inseparable from a contemporary epiphany of an absolutely negative God. Yet I must confess that I feel deeply baffled by being so alone in this apprehension, and above all alone as a theologian, why is it that other theologians appear to be so indifferent or so unresponsive or so unaware of such a contemporary realization of God, all too significantly Barth's original assault upon religion and the religious God has now virtually disappeared, and theologians appear to have no uneasiness at all in the face of our reactionary religious movements, just as they bring no real opposition to our established conservative world. Hence I have become very anti-theological, or deeply opposed to our contemporary theological world, which I am even willing to identify as a Satanic world, and Satanic in sanctioning an absolutely negative epiphany of God. The truth is that our primary prophets, both in China and in Israel, could identify the established God or sacrality of their worlds as a demonic or repressive or inverted sacrality, one turning justice and holiness upside down, and one which could only be the consequence of an ultimate fall or an ultimate reversal. Yet this prophetic truth is seemingly invisible and unspeakable today, and this despite the fact that repressive religious forces are so powerful in our world, and even powerful in a seemingly secular world, as witness contemporary American politics.

Can we join Blake in understanding our new world as the very Body of Satan? Is this the ultimate source of our profound emptiness and darkness, and could an actual unveiling of this source have any effect upon our world? Or is your world so very different that you cannot respond at this point, or would be impelled to respond in very different terms, and are you now engaged in such a project? I simply have no awareness if any such thinking is occurring in this country, our periodicals and journals are seemingly closed to such problems, and now critical thinking virtually never appears in anything but the most esoteric circles. But this is not a night in which all cows are black, genuine naming has not wholly disappeared, and if my vocation is the naming of Satan, I sense that many others have comparable or parallel vocations, and could this be true of you, too?

Hoping to hear from you soon.
Tom

Letter 12

To Lissa McCullough (September 7, 2005)

Dear Lissa,

While I was visiting Walter and Nancy Strauss in Cleveland, we spent much time watching the TV coverage of the hurricane,[1] thereby joining Americans as a whole, and reacting with horror not only to the immense suffering and damage but also to the federal government itself. This seems to be shared by the majority of the American people, and perhaps the time is finally at hand to ultimately challenge this administration. But this only impels me all the more to renew or far rather to wholly revise my own theological language insofar as it is an ethical language or insofar as it can become a radical ethical language. By accident at this very time I was rereading one of my favorite books, Christopher Hill's *The World Turned Upside Down: Radical Ideas during the English Revolution.*[2] Not only are these unbelievably exciting "ideas," but they had an immense impact, ultimately transforming world history, for this was the initial expression of what finally became a world revolution. And this was a deeply and radically Christian revolution, but it is a world that has never been entered by subsequent theology, except insofar as this has all too slightly occurred among us theological radicals. Never did I encounter the English Revolution when I was studying theology, not a word was said or written about it by theologians, as this seemingly became an absolutely forbidden subject. Nor did I ever hear of Gerrard Winstanley[3] among theologians, but at least until Nietzsche, he is our greatest ultimately radical theologian, and greatest radical preacher as well.

Who could even imagine a radical preacher today? Yet I rejoice in the Bush administration, and for the open theological role and identity of this administration, everything is blatantly clear, the ultimate policy is enhancing the power of private property, and absolutely enhancing it, even if this only benefits a tiny minority, and simultaneously reducing as much as possible benefits to the majority, and above all reducing benefits to the poor, and all of this in the name of the Christian God, who here is reverenced by a pure obedience to that One whom

Blake named as Satan. I do not think that any previous administration in our history has so fully and so purely employed such a theological sanction, nor do I think that any previous administration has been so purely conservative or reactionary, and here these seemingly polar dimensions are truly and essentially integrated with each other. But most startlingly of all, there is no real or actual opposition to this administration, and above all not from the Democratic Party, which is simply hypnotized by its opponent, and virtually silent on all fundamental issues. I am fully persuaded that theology is now desperately needed, but this is seemingly unknown to all, and most unknown to our theologians who proceed as though they are facing no ultimate challenge. Yet here I must confess my own guilt, for my writing gives far too little attention to this ethical and political challenge, and while I could plead that no such politics was manifest while I was doing the great bulk of my writing, a small core of my writing does address this issue, and I now feel shame that this is such a small core.

I am primarily referring, of course, to *History as Apocalypse*,[4] that major book of mine that has had the least effect, and is probably the least read. While this theme is not muted in *Total Presence*,[5] this little book is strangely unheard within this arena, and perhaps unhearable. Now so far as I know there simply is no comparable theological work in our time, and certainly not liberation theology, which is finally conservative, nor the theology of hope, which is even more conservative. A fundamental question that I have asked throughout my theological life is whether theology as an ultimate necessity must be conservative. Every modern radical thinker is deeply persuaded of this, and if only for this reason one must rejoice in the radical theological thinking that occurred during the English Revolution, and most profoundly and most gloriously occurred in Milton himself. Yet how revealing it is that theology has closed itself to Milton, too, and this despite the overwhelmingly original power of his *On Christian Doctrine*,[6] a book that the theological world not only ignores but will not even keep in print.

In short I feel impelled at this time to take up and to far more fully take up the challenge of doing a truly contemporary and truly radical ethical and political theology. In one sense this would be a fulfillment of a muted and even marginalized theme of my own work, and the more I think about it the more that this appears to be true, and have I become so hypnotized by nihilism in my later work as to forfeit this opportunity? True, I have thought that nihilism is a way into our contemporary ethical and political world, and that at bottom our new political world is a nihilistic world, and just as this impelled me to write that failed essay for "The Nation," this very failure is perhaps a sign that this is not an effective way to proceed. No, I need a far more direct way, after all almost no one understands nihilism, just as all too few understand either Nietzsche or Heidegger at this absolutely fundamental point. However, I do believe that a language of absolute evil is truly necessary here, and necessary if only to expose the pure hypocrisy of the political world today, for I am persuaded that there is a depth or purity of hypocrisy here that is truly new, and new insofar as it occurs so spontaneously and even "innocently," as witness our much beloved President.[7] Our political analysts are simply not up to such a challenge, nor are our political contenders, and never have such contenders been in such deep need of genuine theological assistance, which they will certainly not receive from the established theological world. The truth of the matter is that it is only the theological mind that can genuinely explore or understand evil, and if I have given most of my energy since my retirement to such an exploration, the time is at hand to concretely embody this exploration, and to do so in such a way that it can be understood by the many instead of the few.

I recall once again those exciting days when I was a television preacher, for that is what at bottom I was when I was continually on television, and then I did communicate with a vast audience, and if nothing else I genuinely offended them, which we both know is an ultimate mode of theological communication.[8] While I fully recognize that I shall never gain such an audience again, I refuse to give up the possibility of addressing a general audience, and think that this could most effectively be done in the ethical and political arena. Perhaps only here is there a genuine yearning for theology today, a yearning inseparable from that theological void so incarnate here, a void now inseparable from that evangelical cacophony now engulfing us, and I must confess that I would love to engage in a debate with an evangelical, as long ago I almost did with Billy Graham. And there is a little known truth about the evangelical mind today, one distinguishing it from its predecessors, and that is that it is not open to an ultimate or absolute evil, as witness its refusal to speak about damnation, or actually to speak about Satan. Here, I distinguish the evangelical mind from the popular evangelical preacher, the latter being truly mindless, and more mindless than it has ever previously been, which is essential to its power.

Indeed, popular theological language is so mindless today that this is a fundamental obstacle to the possibility of doing genuine theology in the public realm, and it is fascinating to contrast this situation with the world of the English Revolution, when the very opposite is true, and genuine theology abounded as it has never before or since. Is there a single Church historian who knows this? Yet it is known by all of the major historians of the English Revolution, and one can encounter far more theology in seventeenth century historiography than one can in contemporary theology, but if ours is a truly empty world theologically, this could make possible an ultimate theological innovation. We should also recognize that there is something truly evil in our theological condition, there is not simply a void that we here confront, but far rather a negative or lacerating void, and one consuming everything in its midst, as witness the consequences of employing theological language in our public world. Such language can only be heard as a curse to the sensitive ear, and a curse in a fully theological sense, one that once could be known as the curse of damnation, and now can be known as an ultimate apostasy, or an ultimate refusal of that which it seemingly evokes. So it is that the very name of God can be known as a curse today, and while earlier this was true in the literary language of a Joyce or a Kafka, now it is true in the public language of theology, or in that theological language that dominates our public institutions and life. Hence that theological language that now plays such a decisive role in the political arena is a truly demonic language, one enslaving rather than liberating its hearer, and one inseparable from that repression that is inseparable from it, but now a repression occurring universally throughout our world, and for the first time in history a universal but invisible and unnamable repression.

This is a fundamental reason why Marxist and Freudian language has become unreal in our world, and that our most sophisticated languages today have no point of contact with our public world, but this is a truly unique situation historically, just as the repression in our world is truly unique. And it is unique because its source or sources are so nebulous, it will not do simply to name the corporate or the fiscal worlds, for while they may well be genuinely evil they are simply not that powerful, nor is it possible to find any manifest source of our impotence or repression, here our philosophy is as empty as our theology, but this is nonetheless a theological rather than a philosophical question, and is so because it is inseparable from the question of absolute evil. Absolute evil is seemingly both unthinkable and unspeakable, but it was profoundly thought by German Idealism, just as it is profoundly evoked in the uniquely Christian epic, so that it is absolutely central in the epic enactments of Dante, Milton, and Blake, enactments of absolute evil that are inseparable from enactments of absolute redemption, so that here there truly is a marriage or genuinely dialectical relationship between Heaven and Hell. Only the dialectical philosophy born in German Idealism can comprehend that genuine relationship, but this is a philosophy as theological as it is philosophical, and most decisively theological in its very understanding of absolute evil, an absolute evil only fully embodied in Godhead itself, but it is the self-saving or the self-negation of the Godhead that is the ultimate source of energy and life. The overwhelming problem here is to translate such a vision or such an understanding into a common language, and into a common theological language, or a language that can be understood by everyone, or the "Here Comes Everybody" of *Finnegans Wake*.

One way to do this is to center upon the question of impotence or repression, and to see that our public theological language is illuminating here, for now "God" can clearly be understood as the source of our repression, as the very speakers of God can no longer speak of God without actually evoking our repression, and evoking the ultimate power and ubiquity of that repression. It is this very ubiquity that is unique today, a ubiquity that we can only associate with God, and whereas once this could only be known by a few, now it can be known by everyone, and by everyone who actually hears our public theological language. Now it is not accidental that God has now disappeared from our theological language, or from our critical theological language, as nothing is now more damning or more self-destructive than actually hearing God, or hearing that God who is now spoken in a hearable theological language. This, too, is historically unique, but it opens up a whole new theological path, and just as once we could know God as the God of absolute judgment, now we can know an absolute or an absolutely ubiquitous repression whose only source that we can name is God, and whereas once the way to God was through an absolutely guilty conscience, now the only manifest way to God is through an absolute impotence or repression, or that repression that only now is all in all. Now just as Marxist and Freudian language can no longer name our repression, is a new world now dawning for theology, for even as theology has always been our deepest naming of evil, is a new absolute evil now actual and manifest that only can be named by theology, and named by that theology which just thereby will speak our actual condition?

Tom

Letter 13

To Lissa McCullough (October 13, 2005)

Dear Lissa,

I much appreciated your prodding yesterday evening. You are quite right to question my attempt to go by way of investigating a total repression, for I must proceed far more directly, and do so once again by thinking of God. But I cannot agree with your remarks about our contemporary secular and religious worlds; I think that neither of them is open to the negative abyss of God, neither of them can face a contemporary *horror religiosus*, and they are in harmony with each other in being closed to our abyss. Unless in speaking of the secular world you are speaking of a very few, and that is literally a very few, as witness an organization like the American Academy of Religion. On the other hand, I am no longer content with writing for a very few, if I am going to make a new turn, I must turn away from that path, and inasmuch as genuine literature continues to have a large audience, I do not see why this should be impossible for theology.

I must now address far more concretely than I have our contemporary silence about God, deeply contrasting this with the cacophony of contemporary fundamentalism, but I sense that these are nevertheless genuinely in correlation with each other, and that the silence of the one is simply the other side of the noise of the other, each making possible a deep passivity in response to the emptiness of our world. That passivity is seldom investigated in depth, and so far as I know has not been investigated at all theologically, but it can be embodied in a silence about God, a silence stilling every possible openness to the abyss of God, and here once again I am profoundly grateful to Bergman,[1] whose enactment of a pure silence has so forcefully conjoined our silence with this abyss. It is not accidental that there is so little serious theological investigation of film, for this would inevitably call forth such an abyss, but so, too, is the bulk of our film criticism closed to abyss, and thus closed to the depths of film itself. Perhaps I should now return to the problem of silence. Silence is primary in *The Self-Embodiment of God;* here and elsewhere in my work an original self-negation of an absolute and total silence is the genesis of God, a genesis that is the origin of speech itself, and a speech unreal apart from the negation or self-negation of silence.

But that self-negation is the origin of an absolute abyss, an absolute abyss absolutely unreal in an original totality or an original silence, but once it has been realized an original silence disappears, or disappears within this horizon, so that within this horizon we can still abyss itself only by passivity and withdrawal, a passivity ever enlarging itself in response to the expansion and finally explosion of absolute abyss. And as we know it is the death of God that is that explosion, an explosion of absolute abyss, one fully embodied in the dead body of God, so after that explosion God can only be heard as that abyss, and we hear that abyss insofar as we are now open to genuine hearing. Hence the necessity of a new and comprehensive silence, a silence that is a genuine witness to this abyss, hence a genuine witness to a uniquely contemporary abyss of God, one that is a true *horror religiosus,* thence demanding the deepest possible passivity, and a true and all too literal silence, even if that silence can become cacophony itself in our religious worlds. Now I am persuaded that only a very few are open to this, but we can become open to it insofar as we can actually hear, and can actually hear our own deeper silence, and hear it as a hearing of the absolute abyss of God.

You will remember my persuasion that an opening to this absolute abyss is born in the prophetic revolution of Israel, only now is the voice of God heard as the voice of abyss, thus making possible the advent of a revolutionary hearing, and a revolutionary hearing that will turn the world upside down. Such a revolutionary hearing is reborn again and again, and it occurs in a pure form in Blake,[2] perhaps our most prophetically modern poet and visionary, a prophecy not only first enacting the death of God but decisively naming God as Satan. Now it is fascinating that Blake scholarship and criticism can so resist this, despite its continual enactment in Blake's mature work, but I take this as an expression of a comprehensive refusal to open oneself to that abyss of God that is a consequence of the death of God. This I identify as perhaps our deepest pathology, for we cannot be so closed to our deeper actuality apart from an ultimate withdrawal or passivity, a passivity that Nietzsche named as *ressentiment,* and a *ressentiment* not namable apart from the full realization of the death of God. If I have been alone in correlating Blake and Nietzsche, this is in large measure because our world has become so closed to abyss, or closed to its full actuality, so I do believe that it is now necessary to open the eyes and ears of our world to this ultimate abyss, and that this is a genuinely theological calling.

You might remark that I have been doing this all along, but I have not done it in a common language, or in a homiletic language, except perhaps in the days of my TV and mass preaching, and are those days now calling me once again? No, that is impossible, and I must take Mark's[3] judgment very seriously that it is not possible for us to write for a truly large audience, and he has attempted this with a depth that is perhaps unique. Nevertheless, can I write for an audience comparable to that tiny congregation for whom I preached when I was a theological student in Chicago? Despite its size this was a truly diverse group, and I sensed genuine communication with them, is that an illusion or is it impossible today? True, I then spoke within the horizon of an orthodox and church theology, one wholly alien to me now, but just as I am persuaded that an orthodox theology can now only address the impotence and passivity of our world, a radical theology should be capable of turning the world upside down at this crucial point, and doing so in its very language or speech. Perhaps it was only in writing my memoir[4] that I became genuinely persuaded that theology by necessity must employ a common language, or a language potentially understandable by all, and it is just its homiletic dimension that most distinguishes theology from philosophy, although here, too, Nietzsche is clearly a theologian.

So the problem comes down to speaking of God today in a common language, and thereby speaking of that ultimate abyss of God that is now our deeper ground or groundlessness, this has certainly occurred in our art as witness Van Gogh and Bergman, vast numbers of people can respond to this, so why couldn't more than a tiny number respond to an open theological language now calling forth this abyss? If our silence about God and our religious cacophony are truly conjoined, why cannot this be openly affirmed, thereby turning silence into noise and noise into silence, for if such a conjunction could actually be heard, it could surely have some such effect, or is this impossible? Frankly. I am now far more interested in addressing this silence rather than this noise, and I deem it crucial that this silence must now speak, and speak its most actual ground, and if theologically that ground is only namable as an ultimate abyss, then such utterance must now become actual if our passivity is not to become total. Yes, let it even become actual through the name of "Satan," it is surely time for such a world to become open to that name, for it is ignorant of it now, as witness our common and even critical meanings of Satan. This alone is a witness to our withdrawal and retreat, one most manifest to me in the pervasive evasion of the Christian epic, wherein the deepest and most comprehensive enactments of Satan occurs, enactments evaded not only by Blake scholars but by Joyce scholars as well. And all too significantly Satan is virtually absent from all modern theology, just as an understanding of absolute evil is only present philosophically in Schelling and Hegel and Nietzsche, unless this indirectly occurs in Heidegger, too.

Can we truly speak of evil today, can we be open to its actuality, or is a Bush language about evil[5] our only actual one, and only actual in its very emptiness? It is fascinating that the only twentieth century philosopher who has spoken seriously about evil is Levinas, and that only in an abbreviated form, but then it was not until German Idealism that serious philosophical thinking about evil occurred. Is the question of evil as forbidden today as is the question of God? And is this, too, an ultimate retreat or evasion, and one inseparable from our new and all too comprehensive passivity? Now it was not until Spinoza that a philosophical thinking of passivity occurred, and then it occurred only in the drawing forth of the absolute necessity of God, a necessity that is a total action or activity, and thereby the very opposite of a pure passivity.

Spinoza could know passivity as the ground of every teleology, and thereby of every previous understanding of God, a tradition that for the first time is deconstructed by Spinoza, and deconstructed in the absolutely new totality of Spinoza's understanding of God. Does our new silence about God parallel this understanding, and parallel it by inverting or reversing it, and doing so by the very comprehensiveness or totality of our silence?

Why is it so difficult to investigate this silence theologically? Is silence itself now unspeakable, and unspeakable as it never was before if only because of its new universality, a universality foreclosing the possibility of an actual calling forth of silence? Now it is not insignificant that there has occurred in our time a new philosophical evocation of the absolutely primordial, an absolutely primordial evoking the absolutely silent, but nonetheless making its silence totally unspeakable, and unspeakable in its very purity as the absolutely primordial. Many identify such thinking as the "turn of philosophy to religion," and surely the movement of return is ultimate in this thinking, and even the movement of eternal return. Could such a movement of absolute return be impelled by the absolute abyss embodied in our condition, a new abyss making all forward movement impossible, and thereby making possible a new and absolute movement of return? So it is possible that theology is far more alive in our world than we can imagine, but is it therein alive only in its most reactionary form, perhaps one paralleling the birth of Neoplatonism in the Hellenistic world? We must remember that the movement of eternal return was not thought philosophically until the advent of the Hellenistic world, then it was the very center of Neoplatonism, a Neoplatonism being reborn today, and even reborn in our purest thinkers, as though it is only an absolute return that can truly or purely be thought today!

We must never forget that Kierkegaard understood the movement of recollection or return as being purely pagan, one wholly other than the uniquely Christian movement of repetition, a repetition that is a forward rather than a backward movement, even though both recollection and repetition are movements to eternity. How odd that philosophical movements to eternity should be occurring in our time, but is the name of God not actually absent in these movements, or absent as anything that we can hear as the name of God? Such a condition is far from all previous Neoplatonisms, but these Neoplatonisms were truly theological Neoplatonisms, and theological as our Neoplatonisms cannot be, unless this is true of Levinas. So is not this too a genuine silence about God, one reinforcing the rule that it is now truly impossible to think about God, and truly impossible to speak the name of God except in a cacophonous language or speech? Surely this is a unique historical condition, but does it make possible a genuinely new theology, and a truly new theological language? That may well be impossible, but only its exploration could establish this, and an exploration that may well be occurring even now.

Tom

Letter 14

To Edith Wyschogrod* (December 2, 2005)

Dear Edith,

I am grateful for the papers that you sent, and most grateful to discover so much mind and energy here; unlike me you show no signs of aging. My only regret is that you seem here to be responding to the requests of others rather than writing wholly from the center of your own work, but since this is something that I never really have done, I can't responsibly assess its significance even if I suspect that it is here a block.

Once again encountering your dialogue with Levinas, I am becoming clearer about my own opposition to Levinas, as here occurs when you speak of the primordial act of expiation, for I wonder whether or not there is any actual disclosure in a Leveniasian biblical hermeneutic, or only a deepening or perhaps purifying of biblical mystery. I hear Levinas as always transforming virtue or morality into a purer form of what Hegel knew as abstract Spirit, for it becomes so abstract as to have no concrete or actual meaning whatsoever, hence it effectively removes us from all actuality, and does so above all in its call for a return to the preprimordial. Infinite, which I can only understand as being absolutely antibiblical.

And I must confess that not even you can excite me about the late Derrida's understanding of religion, it is too bad that he didn't have a censor to prevent him from writing so much or so compulsively, although I can respond to this as a witness to the "Bad Infinite."[1] Since the "Bad Infinite" now reigns more decisively and more comprehensively than ever before this could be an important witness, but it must be interpreted theologically, which I don't see in your analysis here. The essay here that most intrigues me is "Incursions of Evil: The Double Bind of Alterity,"[2] although I wish that you had employed Hegel rather than Bateson[3] as your ground, and not been quite so concerned to employ contemporary discourse, which old reactionary that I am seems far weaker than its earlier counterparts (although I do admire you for continuing your dialogue with analytic philosophy). Yet I must ask if this essay genuinely addresses the question of evil, while I only know Janicaud[4] through your analysis I nonetheless see this as a far weaker instrument to open the question of evil than its counterpart in Baudrillard[5], indeed, as you employ his analysis here it seems to dissolve the question of evil. Of course, this is what the great body of philosophy has always done, but I fear that it is now being done most effectively in the analysis of evil in contemporary philosophical discourse insofar as I am aware of it. Certainly such discourse closes the very possibility of an opening to the total presence of the Great Beast or Satan, so that from my perspective it could only be an ally of that "Bad Infinite."

I think that your most powerful essay here is "The Warring Logics of Genocide,"[6] and what is most distinctive about is its full conjunction or fusion of pure analysis and pure ethical thinking, one that I never find in Levinas, or in any of our great philosophers apart from Spinoza and Nietzsche. Now this is the direction that I think that you should now concentrate all of your powers upon, and frankly I would like to see you become freer of Levinas, and while I think that at bottom this is true I think that you are drawn into detours when you employ the language of Levinas, and particularly so when you employ his earlier language such as the *il y a*[7], although perhaps this is something that I simply cannot understand. But does anyone understand it, or anyone apart from the Levinasian cult? True when contrasted with Janicaud's account of techno-discourse, this does appear to be an ethical opening, but could this possibly be an opening to what you speak of here as "the shiver of horror" or "the frission of horror," can that shiver really appear in the context of the command of alterity to refrain from violence? Can Levinas be understood as effecting the deepest possible leap from the horror of the Holocaust, but once that leap has occurred can there be any actual opening to horror, and above all to the horror of the *horror religiosus*?

More than any other twentieth-century thinker, Levinas has silenced or dissolved the death of God, but is an actual primal language possible for us apart or wholly independently of the death of God?

A final question: Is a genocidal logic fully embodied in our world, and in our actual contemporary world? Innumerable people are saying this but they commonly speak only to each other, or only in the language of yet another contemporary cult, and are having no real or actual impact upon our common world, or can do so only when they wholly disguise all reference to a genocidal logic. Do you sense a genuine calling at such a point? And is this any longer possible in the academic world or in any of our institutional worlds, including the worlds of synagogue, church, and mosque?

Yours gratefully,
Tom

Letter 15

To D. G. Leahy (December 29, 2005)

Dear Dave,

I have just finished reading Badiou's splendid book on Paul,[1] and note that the author has absorbed a good deal of biblical scholarship, for I have come to think that you are alone now as a serious thinker who is indifferent or immune to critical and historical scholarship. Frankly I am delighted that this is the first book of Badiou's that I have read, for innocence is needed here, at least for me, and even if I did learn something of Badiou through you, such as the primacy of "event," here his language is a truly theological language, and "event" is clear as the event of Jesus Christ. From this event and from this event alone does universalization and the subjectification of the subject proceed, for this event is the center of centers and the ultimate ground of all actual liberation. For even if genuine universality is devoid of a center, Paul embodies and enacts a truly and absolutely new universalism, so that the subtitle of this book on Paul is "The Foundation of Universalism." Indeed, this is extraordinarily challenging to the theologian, for a good deal of this language is more truly and more actually theological than the great body of language in the theological world, and it is true theological power as such that is realized here, and often realized by astute textual exegesis. And irony of ironies, he even actualizes Nietzsche as an ally of Paul, and a weaker one at that!

What is most startling here is Badiou's language about God, a language missing from most critical theology today, and while it is true that this language is somewhat cryptic, it is far less so than that of Heidegger and Levinas, and even enacted in a language that the theologian can recognize as being genuinely evangelical. Perhaps the real secret of Badiou is that at bottom he is a theologian, and even a Christian or universal theologian,[2] and never before have I encountered such truly theological statements that are enacted not only as a universal call and address, but as a call to a universal body and action. I think that nothing is more fundamental here than Badiou's initiation of his own discourse on God, for as he declares the Christ-event (as understood by Paul) causes nonbeings rather than beings to arise as attesting to God (p. 47), so that God is certainly not Being, and the pure "event" of Jesus Christ is the advent of the subject, one never known or actual before.

Now the singularity of the subject is established and it is established by Paul's own language, and it is established in a third discourse, one going beyond Greek and Jewish discourse, having the rude harshness of public action, without apparel other than its real content, and it will be nothing but what each can see and hear, for this is the earthen vessel. Now Christianity establishes a potential universality, and the subjective element is now rendered adequate to that universality, for this is the initiation of a subject divided in itself by the challenge of having nothing but the vanished event to face up to. Once Protestant theologians emphasized the kerygma and the kerygmatic identity of the language of faith, this is renewed in Badiou by his emphasis upon declaration, and even renewed in his statement that Christ as Son is emblematic of the way in which the evental declaration filiates the declarant. For Badiou the resurrected Son filiates all of humanity, but here he differs from the real Protestant, who knows this to occur already in the Crucified, and Badiou does lessen if not virtually eliminate the Crucifixion, here profoundly differing from Paul. Remembering that for Paul the Crucifixion is a profound offense and scandal to philosophical wisdom, that offense is renewed here, even if Badiou does deeply celebrate the antiphilosophical Paul.

Now for Badiou's Paul Jesus is the name of what happens to us universally, and the event of Christ is the event of grace, and if all that matters in Christ is the Resurrection (Badiou not Paul!), there is a superabundance in this event for all. Badiou's refusal or evasion of the Crucifixion is fascinating, perhaps because he understands how primal this is for Hegel and for the Hegelian dialectic, which he resolutely refuses, and in doing so refuses a dialectical Paul. Yet Badiou remains within the shadow if not the perspective of Hegel in understanding the Cross as the immanentization of Spirit, and through this death we cease to be separated from God, for it is the unique necessity of Christ's death that gives us the means to an equality with God. Through Christ's death God renounces his transcendent separation, and unseparates himself through filiation, and shares in a constitutive dimension of the divided subject, thereby becoming what Badiou calls the site of the event. If only by way of an analysis of Nietzsche, Badiou comes to know the event as bringing about an unqualified affirmation of life against the reign of death and the negative, and ultimately for Badiou's Paul the Christ event is nothing but resurrection, one eradicating negativity, and terminating the abyss of transcendence.

It is in his chapter on Paul against the Law that Badiou most decisively speaks of God, and does so by speaking of Paul's revolutionary conviction that the sign of the One is the "for all" or the "without exception," for the One God inscribes no difference in the subjects to which it addresses itself. The One *is* only insofar as it is for all! Thereby we are given a grace that is the opposite of law, for it comes without being due, and it comes to all. It is precisely in his language about grace that Paul most purely decenters the subject, and here Badiou can even speak of salvation as a deliteralizing of the subject, for "grace" names the event as a condition for "active thought," by which he perhaps means pure action or pure engagement. Then he goes on to discuss love and hope, maintaining that it is incumbent upon love to become law so that truth's postevental universality can continuously inscribe itself in the world, and rally subjects to the path of life. Then in his conclusion Badiou speaks of the Pauline break as having the inevitable consequence of bringing forth a consciousness of truth rooted in a pure event, detached from every objectivist assignation, and concretely destined to become inscribed within world and society. And Paul must be given exclusive credit for establishing that the fidelity to such an event exists only through the termination of communitarian particularisms and the determination of a subject of truth who indistinguishes the One and the "for all."

I have very largely simply reproduced Badiou's language here, and done so simply to make clear how deeply theological it is. Of course, he disappoints both of us by largely if not wholly ignoring the apocalyptic Paul, or transposing that Paul into an image of his own work. Since I have only begun Badiou, I cannot say that, but I do suspect that he is ignorant of the more important Pauline scholarship, and perhaps because of his own French situation, which keeps deeper biblical scholarship in the shadows. I am deeply grateful to Alina for this gift, and she also gave me the Badiou anthology, *Infinite Thought*,[3] which I shall next take up.

Happy New Year,
Tom

Letter 16

To Brian Schroeder*
(January 23, 2006)

Dear Brian,

I am grateful for your remarks on nihilism and for the D'Agostini article,[1] which while presenting the centrality of nihilism in contemporary Italian philosophy has a rather weak analysis of it. He did interest me in Sergio Givone,[2] however, is he in your stable?[3] Lissa has suggested that my new book be on nihilism and this possibility is now intriguing me. I gather that the Italian philosophers are not only atheological but removed from Eastern philosophy as well, and that would surely deeply limit one's understanding of nihilism, and probably make impossible an understanding of its ultimate depth. But I trust that everyone now seriously investigating nihilism must be under the impact of late modern literature, and just as Heidegger's immersion in modern German poetry deeply affected him at his point, any such immersion should make impossible a naïveté about nihilism, which is why I presume that most of our philosophers and theologians are unengaged with nihilism. As you know, I have recently become engaged with the possible power or ultimacy of a popular or mass nihilism, and just as I have long sensed that fundamentalism is genuinely nihilistic, I think that this is a decisive key to our mass media as well.

I am intrigued that so far as I know philosophical investigations of it ignore Nietzsche's claim and even demonstration that Christianity is the origin of nihilism, this claim cannot be grasped without realizing that Nietzsche's understanding of absolute No-saying is inseparable from the uniquely Christian God, and most particularly so as this is understood in *The Antichrist*. So, too, Nietzsche understands that ultimate historical transformations have occurred through or by way of that God, including not only the advent of the bad conscience but that ultimate historical event of the death of God, which made possible Nietzschean thinking itself. Now if this releases a uniquely modern nihilism, and the first absolute or total nihilism, such nihilism is realized by way of the tomb of God, or by way of an absolutely new emptiness or vacuum. Only then did the West become open to Buddhism, which it inevitably understood as a pure nihilism, as witness Nietzsche himself, and even if this is a radical distortion, it nonetheless is inseparable from our new opening to Buddhism.

No doubt it is ultimately important to radically distinguish a uniquely modern void or emptiness from a Buddhist emptiness, yet there are nevertheless genuine parallels between them, as can most clearly be seen in Nietzsche's enactment of Eternal Recurrence, which is an enactment of that absolute No that is an absolute Yes. As you know, I have long believed that this No is the uniquely Christian God, but thereby we are given an apocalyptic Yes, and it certainly is not insignificant that those philosophers who have deeply enacted the death of God, Hegel, Nietzsche, and Heidegger, are our most apocalyptic thinkers. I think that this is the major reason why few philosophers can understand them theologically, for philosophers are perhaps even more closed to apocalypse than are theologians, although Italians may well be an exception to this. Of course, genuine Marxists are grounded in apocalypse, and is this a primary source of Badiou's distinctiveness? I find it fascinating that those French thinkers who moved into an absolutely primordial thinking became so profoundly alienated from apocalypse, and did so with apparently no realization that Christian, Jewish, and Islamic theological orthodoxies were all realized by way of an absolute negation of apocalypse.

Could it be that apocalypticism itself is a primary way into our nihilism, and is so if only because of its ultimate enactment of an absolute ending? Again, Hegel, Nietzsche, and Heidegger each enacted such an ending, yet few of their interpreters can truly enter that ending, and if only thereby a crucial way into nihilism is refused, for in its ultimate form it is a consequence of such an ending. And that ending is a truly historical ending, which is to say that it ultimately impacts upon us all, and thus there is inevitably a popular or mass nihilism, and it is certainly not confined to our former political totalitarianisms. This is also a way into understanding our uniquely contemporary totalitarianism, one that our politicians can only vaguely or indirectly sense, for this is a totalitarianism that is as fully present in the mass media as it is in mass politics, and just as our politics and our media induce nausea whenever openly sensed, there is really no possibility of either evading or hiding from either, and hence no way out for us. That, too, is an apocalyptic situation, and apocalyptic in its very ultimacy, and in the unique ultimacy of an absolute ending.

Tom

Letter 17

To Walter A. Strauss*
(February 6, 2006)

Dear Walter,

It is good to hear from you regularly and I much appreciate your prodding, which helps to keep me alive. Or am I alive? I just completed an essay that I may or may not have sent you, I include it in case I didn't, and there are problems here that I would like to share with you. Remember the myth and ritual school of criticism that Herb Weisinger[1] embodied? This had an effect on me larger than I realized as reflected in this essay, but then the problem of ritual has been with me almost from the beginning, and I have been able to do little with it. Now I am once again taking up the problem of our nihilistic destiny and confronting the concrete problem of just how this destiny is mediated to us. This is the context in which I attempt to take up the question of ritual, realizing that we know very little about it, and perhaps are most ignorant of it theologically.

Suddenly it seemed to make sense to me that there is an invisible ritual or liturgy mediating a pure or absolute nothingness to us, for I don't think that our pervasive nihilism can be accounted for by anything that we can understand as a cultural, social, or historical influence; there is a pure immediacy and universalism in this nihilism, which I sense can only be accounted for by a liturgical mediation. If so, the problem is that we have so little understanding of liturgy, almost wholly associating it with an ecclesiastical domain that is irrelevant to this problem. Now here the myth and ritual school seems all too relevant, and if we can easily see a mythical dimension of literature, why do most miss its ritual dimension? And is a ritual dimension inseparable from a genuinely mythical dimension? Or is all this alien to our criticism? I don't know enough to be sure, but you do. Nothing has more excited me about *Finnegans Wake* than its liturgical mode, one that becomes immediately real in oral readings from the *Wake*, or in the recording of Joyce's reading of a section of it. And it is the liturgical mode which could account for the immense impact of the *Wake*, one almost more real because its text can be so little understood. I am intrigued that its most difficult and I think its most powerful section, which is often understood as the axial chapter of the *Wake*, "How Buckley Shot the Russian General," is a repetition and reversal of the Roman liturgy, and one now liturgically enacting that death of God that is apocalypse itself.

Many see the *Wake* as a comprehensive embodiment of our nihilism, but is it not true that this is our purest and deepest liturgical literary text, and one that enacts itself liturgically when we enact it? What would you site as liturgical texts in our literary canon? Is there a liturgical dimension to Proust or to Dostoyevsky or to Rilke, etc.? What about film? I seem to sense it in film but can't articulate it, and an example that immediately comes to mind is Carl Theodor Dreyer's *Passion of Joan of Arc*[2]. Yet I think that all drama is in a fundamental sense liturgical, and perhaps this is most powerful in late modernity in Beckett. Has ritual or liturgy been employed in our understanding of Beckett? Or Kafka?

In this essay I employ immediacy and universality as being characteristic of liturgy, and one distinguishing ritual from myth, but this is only a crude and all too limited beginning, and one demanding far more exposition. Should I venture upon that now? Or can I? Or is this actually a fruitful path for me to explore? Incidentally, my primary reading now is a re-reading of Harnack's *History of Dogma*, this had an immense impact upon me when I was a graduate student, and I appreciate it now theologically more than ever (surely it embodies our deepest understanding of Augustine!), and so far as I know this is our deepest excursion into ancient Greek or Patristic theology, and despite Harnack's Protestant prejudices, that theology here truly becomes alive, but it is liturgical theology that is least powerful, with the possible exception of Gregory of Nyssa. Harnack is deeply repulsed by the profound reaction which occurred in Eastern Orthodoxy, which for him is little more than a regression into magic, but it is Eastern Orthodoxy which became the purest or most comprehensive liturgical Christianity, and if only thereby has become far removed from Western Christianity and the West as a whole. Or is a deeper Western secularism a renewal of such a way? And is this now embodied in our nihilistic destiny and our nihilistic world?

Yours faithfully,
Tom

Letter 18

To Lissa McCullough (February 7, 2006)

Dear Lissa,

I am most grateful for your inducing me to return to my chapter on primordial sacrifice, I simply did not realize as you did how close this is to my last paper, but now I cannot only see this, but can see just how minimal and fragmentary this chapter is, and there is certainly the potential here for an expansion into a book. Perhaps this is the book that I now must write, and I have long had a sense that I should devote a book to sacrifice, and to absolute sacrifice, for while this has always been my primal subject, I have not focused upon absolute sacrifice as such, or upon absolute sacrifice as not only our deepest call, but a call that we can actually renew because it has not only absolutely occurred, but is even now absolutely occurring. You are not alone in regarding the chapter on primordial sacrifice as the best one in *Godhead and the Nothing*,[1] and while I do not share this view, I have seen it as that chapter of mine with the richest potential for development and expansion, and now this is even clearer, for not only is this a fragmentary statement, but it embodies in a wholly minimal form my deepest theological projects or motifs.

Even now I am rereading Harnack on Augustine,[2] our greatest critical investigation of Augustine, and Harnack does know the ultimate importance of predestination here, but he can't fully or decisively articulate it, and while this motif is radically minimized in the chapter on primordial sacrifice, this is an excellent example of what could come forth if this chapter were to be developed or unfolded. Indeed, and despite all the work I have given to predestination, I have not truly or decisively articulated it as absolute sacrifice, and surely it is only thereby that its radical offense can be absorbed and affirmed. I have also been reencountering Beckett recently, and particularly through Netflix distributed films of his dramas, and when these are realized or understood as expressions of primordial sacrifice, they become all the more powerful. The acting and direction in these films are quite marvelous, but thereby the sacrifice is all the more real, and far more so because the dramas are so absurd or meaningless. Yes, this is pure ritual action, and pure sacrificial action, but that is just why it is so powerful, and most powerful when it *is* truly absurd.

Again, I touch on this in the primordial sacrifice chapter, but just barely touch it, indeed, I seem to do little more than touch anything here, and perhaps this is best for many, because I can spoil ultimately important motifs when I develop them in my own language. Of course, in a crucial section of this chapter I am simply transposing the language of the *Phenomenology*, and while this is made clear here, it is Hegel's language and not my own that is ultimately powerful. As you know, I deem this the most important text in the history of philosophy, and it has generated untold riches, including even in spite of himself Kierkegaard's greatest texts, so I am not ashamed to ally myself with it, but I would like to be able to follow Kierkegaard here and wholly transform this language into my own, or into the text at hand. Something like this occurred in *The Self-Embodiment of God*, and perhaps it could occur again, and occur even now. Dare I pray for such a miracle?

Just evoking *The Self-Embodiment of God* does forcefully remind me that I have unfinished work in developing a biblical theology, and there are fragments on that, too, in this tantalizing chapter, and I am not speaking so much of the New Testament as of the Old, and not the prophetic revolution, but rather the biblical epic itself, which perhaps can best be understood as the realization of absolute sacrifice, one that is ironically reversed in the Book of Job. Auerbach's understanding of *mimesis* is deeply relevant here, and while this is best unfolded in his book on Dante,[3] it purely occurs in *Mimesis*,[4] and while he does not employ sacrificial language to draw it forth, that could certainly be done, and done so as to show how the birth of *mimesis* is an historical actualization of an absolute sacrifice. Here it is Jewish critics and scholars who are most powerful, and perhaps most powerful because they are unaffected by the perspective of the New Testament. And not since the early Barth have we had a theological or critical understanding of the New Testament that draws it forth as a true *mysterium tremendum* or a true *horror religiosus*, and how fascinating that recent philosophical investigations of Paul are wholly innocent of this *horror*, and thus from my perspective are incapable of truly becoming open to Paul. Perhaps only the Christian can do this, and do this because of a unique and absolute Christian guilt, and one inseparable from a uniquely Christian grace. This is just where our exegeses of Augustine are most powerful, and where I believe Augustine is most profoundly alive today. Perhaps the best section of the primordial sacrifice chapter is its drawing forth of the Book of Job as a pure enactment of a *horror religiosus*, and of that *horror* as a pure reflection of the absolutely sovereign and absolute transcendent Creator, but a Creator who is only possible through an absolute and absolutely primordial sacrifice.

Yours gratefully,
Tom

Letter 19

To Friends (March 2, 2006)

Friends,

Ratzinger's *Introduction to Christianity*[1] is now clearly an important theological work for us, and it is far more radical than it is commonly perceived to be, indeed, I am impelled to record my ironic apprehension of the good cardinal or pope as being a "death of God theologian." No one can open this book without immediately becoming aware of the ultimate importance of the year of its publication, 1968, as one is initially given a reprisal of the book by its author written in 2000, calling attention to the year 1968 as a time of ultimate nihilism, when an abyss of nothingness was at hand, and everything changes as Nietzsche's proclamation of the death of God is realized in our midst. This is the context in which this book was written, and one accounting in large measure for its deep impact, and even if Ratzinger can say that he is writing in the spirit of Karl Adam's *Spirit of Catholicism*,[2] a radically new world is here at hand. Not even Rahner was open to our nihilism, and at that time in the world of Catholic theology only Teilhard de Chardin[3] was open to nihilism, and Teilhard has had a real impact upon Ratzinger, but virtually none upon other major Catholic theologians. It is well known that Ratzinger reacted very strongly against 1968, but this is commonly associated with the student radicals of that period, whereas his own book makes it overwhelmingly clear that he was most deeply responding to a new and ultimate nihilism. And to judge by the prefaces of the book, his understanding of nihilism is far deeper in 2000 than in 1968, leading one to surmise that Ratzinger has reflected more deeply on nihilism than any other Catholic theologian, which I suspect is the deepest ground of his conservative and even reactionary theological turn.

It is remarkable how few are aware of the possibility that it is a new and comprehensive nihilism that has been the deeper ground of our new conservative theological world, just as it is also fully possible that it is a reaction against such nihilism, even if only an unconscious one, that is deeply responsible for our new conservative political world or worlds. The truth is that very few among us have a serious understanding of nihilism, and there is no critical agreement today upon the meaning of nihilism, nor is there even a good book on the subject, with the inevitable exception of Book 1 of Nietzsche's *The Will to Power*. Yet Ratzinger does have a truly astute understanding of nihilism, and in *Introduction to Christianity* he can allow this to set the stage for a new investigation of the Christian faith, almost reminding one of the young Luther, and particularly so as Harnack presents him in the final sections of his *History of Dogma*.[4] We are want to forget that Ratzinger is a deeply German theologian, and one who has inevitably been affected by a uniquely German philosophical theology, which is our only truly modern philosophical theology, and perhaps most modern in its ultimate enactment of the death of God. All too significantly many of Rahner's opponents have found such a philosophical atheism in his work, just as conservative Thomists find it everywhere in modern philosophy, as marvelously enacted in Cornelius Fabro's *God in Exile*. Of course, Barth, too, had a deep understanding of modern atheism, and it is not only at this point that Ratzinger can be associated with Barth, but also insofar as each have battled so fiercely with uniquely modern expressions of religion. This is most manifest in Ratzinger's assault upon Liberation Theology, for this is not only an attack upon a Catholic expression of Marxism, but also an attack upon a radically independent form of faith and community, one setting itself against the Hierarchy of the Church, and even against all hierarchy whatsoever. Now it is just in a nihilistic or radically atheistic context that an overwhelming need of hierarchy looms forth, one not only creating the radical hierarchies of modern totalitarianism, but profoundly affecting all institutions, and most clearly so the Catholic Church.

Ratzinger is the first pope who is a professional theologian, or a major theologian, and one who theologically eclipses virtually all of his theological critics, so we can expect to witness gnats battling an elephant, which hopefully will lead the Catholic world to take theology far more seriously than it now does. To judge by his first encyclical we can expect Benedict XVI to be fully responsible theologically as Pope, and there is a particular power here insofar as this is a theological response to an abyss of nihilism, one that John Paul II attempted to articulate, but was unable to do so. But given our comprehensive ignorance of nihilism, we cannot expect astute responses to this papacy, or even an appreciation of an ultimate battle that may well now be waged. One is reminded of a story that Ratzinger tells in the opening of *Introduction to Christianity*: a traveling circus in Denmark catches fire, the manager sends its clown into a neighboring village, where he screams for help, but the villagers think that this is only a circus performance to lure them to the show, and despite the clown's weeping they can only laugh at him, until finally the fire engulfs the village, and everyone perishes. Obviously, Ratzinger associates himself with this clown, and this could even be an image of his papacy, for there is an apocalyptic sensibility in this pope that is rare, and also an astute sense of an absolute abyss looming within and beneath us, and one that inevitably must explode, so that the deepest duty of the Church is now to prepare for that explosion, and to provide sustenance against it. Ratzinger is known for preferring a far smaller but far stronger Church, here one can sense the impact of Kierkegaard, but his commitment to orthodoxy is not that common one that appears about us, it is far rather a commitment to a new orthodoxy, and one occasioned only by an absolute crisis.

Perhaps the clearest analogy to this crisis is the Augustinian crisis in the ancient world, then a world was clearly coming to an end, and one demanding a truly new theology, Augustine created such a theology, and thus made possible not only Western Christian theology but a Western Christian world. Now we can expect no such miracle from Ratzinger, but is it possible that he is now in position to initiate such a transformation, and even to give it an institutional foundation? For there surely has been no previous pope with such a sense of an ultimate crisis that is a theological crisis, one profoundly uprooting us in our deepest foundations, and therefore ultimately challenging our deepest faith. At the beginning of his first Preface to *Introduction to Christianity*, Ratzinger remarks that the Christian faith is enveloped today in a greater fog than at almost any earlier period of history, and this not simply because of a comprehensive watering down of faith, but because of a contemporary faith hiding a complete spiritual vacuum. Apparently no other Catholic theologian has such a deep sense of this vacuum, so that in his opening discussion in this book on belief in the world of today, he can speak not only of the power now of unbelief, but of the bottomless abyss of nothingness that now can be seen, and even seen as he observes by our saints themselves. This can induce him to initiate an apparently new theology, one in which God is *essentially* invisible, and while he speaks as though this is simply biblical theology, the biblical world or worlds surely did not know a bottomless abyss that is an incarnate abyss, and thus inevitably drawing forth a truly new or absolute invisibility of God. This is very close to Kierkegaardian language, and truly distant from traditional Catholic language, to say nothing of the whole world of natural theology. True that world has long since perished, but Ratzinger knows this all too well, and knows it as no previous pope has known it, to say nothing of the common forms of Catholic theology. So Ratzinger may yet prove to be a revolutionary pope, even if he is revolutionary by establishing a truly new Catholic orthodoxy, an orthodoxy inseparable from an ultimate and comprehensive nihilism, and only made possible by that nihilism itself. Both Kierkegaard and Dostoyevsky have given us such an orthodoxy, but none such has yet arisen in the Catholic world, can we realistically expect such a gift from this papacy?

Thomas J. J. Altizer

Letter 20

To Lissa McCullough (March 9, 2006)

Dear Lissa,

Once again although this letter is addressed to you, it is really written to myself in an attempt to record in small part what occurred when I just re-read *Genesis and Apocalypse.*[1] You primarily remember here the rethinking of Augustine and Nietzsche so as to bring forth a new and deeper meaning of predestination, and even to call forth Nietzsche and Augustine as one thinker, but this now least affects me if only because I have so fully absorbed it, although I do regret that it has had no affect upon Nietzsche or Augustine scholarship. So far as I know Augustinian scholarship is now so weak that it doesn't fundamentally rethink anything, and Nietzsche scholarship although very strong is not open to rethinking Nietzsche in another perspective or horizon, which my work demands. Indeed, Nietzsche scholarship almost wholly refuses Nietzsche as a fundamental theological thinker, despite its pervasive presence throughout most of his work, which I think is largely because it can only think of theology as a church dogmatics, and while that is a responsible reaction to contemporary theology, it does entail a refusal or evasion of the profound theological thinking that occurs throughout German philosophy.

No, what most affected me in this rereading is the pervasive role of nothingness and an absolute nothingness throughout this work, just as nihilism is central here in a way that I had forgotten, and central as a constructive although largely hidden force. I immediately recognized once again, and this I did know long before this, that the first two chapters are only all too partially worked through, and the third chapter is so weak as to demand a total revision, and the problem is so major here that I have real trouble in attempting to re-read these chapters. Yet there is a genuine potential in the first two chapters, one in part realized in later work, but far more can and should be done, and perhaps will be done in the coming year. Yesterday I was reading Bernard McGinn[2] on Eckhart's understanding of creation, a reading impelling me to return once again to the deepest problem of these opening chapters, for just as Eckhart can only know an eternal creation, that very perspective gives a new urgency to understanding creation as an absolutely new creation, and an absolutely new creation that is the consequence of what Eckhart understands as the negation of negation, although this demands a pure inversion of what Eckhart knows as that One that is the Nothing. Raising such questions clearly evokes an ultimate taboo, for we are deeply forbidden to think about a precreation, and I love that apocryphal story that when Augustine was asked what God was doing before the creation, he replied that he was preparing a Hell for those who asked such questions.

I now think that the chapter on "Emptiness and Self-Emptiness" in *Genesis and Apocalypse* embodies my best work on Buddhism, and this identification of Emptiness and the primordial Totality is potentially extraordinarily important theologically, and it certainly could lead to a truly new understanding of apophatic theology, just as it impels a radically new understanding of creation. If only as a thought experiment, the idea that the nothing out of which the world is created is Nirvana itself, thus calling forth creation as fall in the most ultimate sense, certainly opens immense theological possibilities, and one of these is that it is an original totality or bliss that is our deepest and most destructive or pathological temptation, and one openly embodied in Gnosticism. But Gnosticism is only possible as a rebellion against the biblical God, hence it is unknown in Buddhism or Hinduism, and possible only under the impact of Judaism, Christianity, or Islam. While I am one of the few who insists that Gnosticism did not and could not exist before Christianity, for Gnosticism originates in Christianity, the first witness to it is the New Testament, and what we know as ancient Gnosticism is profoundly Christian, and arises as an absolute rebellion against the uniquely Christian Creator, a Creator who is Creator and only Creator, and only thereby can be named as Antichrist or Satan. Harnack captures this powerfully in his book on Marcionism,[3] and just as it is Harnack who most decisively unveils ancient Christian heresy as being profoundly Christian, it is heterodoxy in our world that is far more profound than orthodoxy, as witness a Blake or a Hegel. Barth himself is far more profound as an heretical rather than as an orthodox thinker, as witness his primal doctrine of election or predestination, which is certainly not a church dogmatics, and could not be in its ultimate claim that only Christ actually suffers damnation, and it is that damnation that is the Gospel itself!

I now rejoice more than before in the Barth exegesis in *Genesis and Apocalypse*, and for its conjunction of Barth and the Milton of *Paradise Lost*, and only this analysis makes possible the chapter on Christ and Satan. There is a doctrinal or dogmatic comprehensiveness in G & A that goes beyond my other writing, it is not for nothing that this book was so shaped by Barth's *Church Dogmatics*, but it is certainly attempting to turn Barth upside down, and I now think that it succeeds in this venture. What is exciting here is to wholly reverse a pure orthodoxy, I think that it is possible to understand that a pure apophatic theology does this in its own way, hence we can understand Erigena[4] as the father of Western heterodoxy, and even apprehend a genuine continuity between Erigena and Blake and Hegel. It is not for nothing that our Church historians since Harnack refuse to investigate heterodoxy, or become so consumed by it that orthodoxy itself is forgotten as in our Gnostic scholarship, and if this has led to the virtual collapse of Church history as a serious discipline, this is yet another sign of the profoundly pathological condition of the Church today. Ray Hart affirms that unless we are a Kierkegaard or a Nietzsche we can only survive or have impact through the Church as an institution, but he insists that the true or actual Church is a radically reforming Church, one that I can see no signs of whatsoever in the contemporary world, and I certainly mean to include the contemporary theological world. But could my theological work be included within a radically reforming Church, and is such a church even an institutional reality in our world, and if so how could we recognize it?

Rereading this book impels me once again to think about theology as systematic theology, and systematic in a sense I never knew as a theological student or theological novice, and if an Eckhart is most systematic in his thinking about the One or the Nothing, just as Hegel is most systematic in thinking about an absolute self-negation or an absolute self-emptying, then in an infinitely lesser way this book is most systematic in its thinking about nothingness and an absolute nothingness, although it fails explicitly or actually to do this throughout most of the book. I recognize that impatience is perhaps my worst fault, and that I should have waited far longer before writing this book, but I can become possessed by a theological passion when I am driven to write a book, and I was possessed when I wrote this book, which is probably obvious to a serious reader. But can we or do we understand a theological obsession or possession? Is possession inevitably and necessarily a demonic possession, I refuse to think so if only because I know both Kierkegaard and Nietzsche as being possessed, to say nothing of Blake, and here possession releases the greatest possible creativity, so that my problem is can a far lesser possession release at least a fragment of genuine creativity? Now when I read this book of mine today, I recognize that its best fragments are not only far from being my own, but are fragments that I now cannot genuinely understand, and insofar as I do understand them they stand forth as being wholly potential rather than actual, and even if this now offers me the possibility of genuine work, it does so in such a way as to arose a deep *Angst* that such work is no longer possible for me, and that is an encounter with even a literal nothingness.

As I was doing this rereading I would often ask myself if I could revise a given passage of it today so as to make it more real, and except for the dreadful third chapter I feared that I could not do this, unless I simply began all over again with the first three chapters. This is now beckoning me as a possibility, but it would demand a genuine rethinking of this writing, and one making it far more systematic, and far more systematic in its evocation and celebration of an absolute nothingness, or the absolute inversion and reversal of an absolute nothingness. As I think about this, I feel the temptation to become obsessed once again, if that is any longer possible, but now openly possessed by a systematic thinking about nothingness and about nothingness alone. Could Eckhart have been a genuinely systematic thinker even if he left no actually systematic writing, I can accept this if only because of my harsh reaction against all of those who refuse to recognize Nietzsche as a systematic thinker, or his corpus as a finally systematic body of texts, and certainly an affirmation of Nietzsche as a truly systematic thinker brings a new meaning to systematic thinking, which so many can think of as only occurring in a Hegel or an Aquinas, and occurring here in such a way as to establish the paradigm of systematic thinking itself. Buddhists have taught me that Nagarjuna[5] is a profoundly systematic thinker, and is so most purely by thinking about "nothingness" alone, and it is truly remarkable and ultimately important that it is Buddhism that first evolves or creates a truly systematic thinking, and an even more systematic thinking than the Aristotle corpus could produce.

This alone can give us a new and more responsible understanding of systematic thinking, and when I think back upon it I realize that the very project of writing *Genesis and Apocalypse* inevitably entails a fundamentally systematic thinking, for it demands a thinking of genesis and apocalypse at once, a conjunction that presumably had never occurred before, and this is a conjunction that can only occur by way of a systematic thinking, for to think genesis as apocalypse is to think not only of the final consequence of genesis, but of an apocalypse that is realized only through genesis itself. Of course, Hegel's *Science of Logic* thinks genesis and apocalypse systematically and purely systematically, but here apocalypse is wholly veiled or disguised, which is perhaps necessary for the ultimate profundity of that absolute genesis that here occurs. And why is so little attention given to Hegel and Schelling as the first Western philosophers of nothingness and an absolute nothingness, or the first post-mystical thinkers of nothingness, and why is this not inevitably associated with that Hegel who is the first philosopher of the death of God? I sense that philosophers here, and above all Heidegger, are silent about this because they are so deeply committed to a silence about theology, and even if Heidegger is a profoundly theological thinker, he can only be so by deeply disguising his theological language. I often think that the deepest hostility is directed against me because I am so reckless or uninhibited theologically, maybe this is truly a product of the Sixties, and is unreal apart from that, or apart from a comparable ground. And even if we are all now shaped by a world placing a heavy emphasis upon thought experiments, a deep lacuna or taboo has been placed upon this theologically, as though it is theological language alone with which we dare not experiment, and this is subtly but powerfully enforced even in wholly non-theological worlds.

Now when I read the Prologue to *Genesis and Apocalypse* a few days ago I could certainly sense that something truly systematic is occurring here, now I have just re-read it again, and discovered to my amazement that "system" and "nothingness" or nihilism are truly and openly conjoined here, and it is only in that context that "system" here has any meaning at all. How could I forget this so quickly? Of course, memory is a major problem for me now, but it is rare for such a lapse to occur, and I cannot avoid suspecting that a genuine repression has here occurred. "System" and "Nihilism," are these fundamental words inevitably conjoined for us, and if we think carefully about it are not Hegel and Aquinas those very thinkers who are most condemned in our world as a whole? I have long been persuaded that nothing is more revelatory than that which is most deeply condemned, and above all that which is most condemned in one's own world, and in my world that which is most condemned is any open or actual language about God, a deep iconoclasm is operative here, and even operative in our most secular circles. Now that is the condemnation that I have most openly defied, and I recognize that this did not occur until the Sixties, and I undoubtedly was affected by the radicalism of the Sixties, a radicalism deeply liberating me, and one that I can now recognize as an ultimate gift. It interests me that in virtually all of the literature that I have read on the Sixties there is no mention of theological radicalism, but this was a time when the death of God controversy for a brief period possessed the mass media, and in which innumerable religious figures were truly radical actors, I know if only because I lived through it that a religious or theological radicalism was truly real in a great many radicals, and even to this day atheism and radicalism are identified in our common language. Has all of this simply been forgotten? In our extraordinarily conservative or reactionary world, the Sixties are virtually always spoken of negatively, it is as though the great art, music, and literature of that period simply doesn't exist, only its rock stars are now revered, but rock has been defanged in our world, just as have all of our cultural expressions.

I must confess that I very much want to believe that it is theology that is most condemned in our world, nothing in my judgment could make theology more important, or more expose the emptiness of the theological world today, but far more significantly for me, this opens my own theological work to myself in a liberating way, for perhaps my only real accomplishment, or my only real gift, is simply doing theology, and doing theology by thinking about God, which in our world could only mean a radical thinking about God. Notice how Heidegger, who is so extraordinarily careful about language, virtually never speaks of God, and openly does so only in his posthumously published *Beitrage*,[6] and yet it is now all too clear that Heidegger is a profound theological thinker.

Something like this is even true of Whitehead, and of Wittgenstein and Husserl, too, so that in the twentieth century and the twentieth century alone there is virtually no philosophical language about God, or major philosophical language about God, or philosophical language about God that has any point of contact with our theological language. So, too, there is no language about God in our hermeneutics, thereby radically distinguishing our hermeneutics from all earlier hermeneutics, and this absence is scarcely noted in our hermeneutical worlds, it is almost as though the very word "God" no longer exists. Indeed, a world of biblical scholarship and hermeneutics has now arrived in which there is virtually no language about God, and seemingly no theological awareness whatsoever, so fundamentalists can now participate in this world without any sense of incongruity. Many can greet such a world as a truly liberated world, one free of that theological ground that provoked such violence and catastrophe in our history, and when a truly religious world suddenly arises before us, as is true of Islam today, an ultimate shock occurs, and no one knows how to respond to it. While it is true that there is now an increasing awareness that religion is humanly inescapable, there is virtually no awareness that this is an ultimate challenge to us, and an ultimate challenge to our thinking today.

The preface to *Genesis and Apocalypse* declares that this book intends to bring together the thinking of my earlier books, *The Self-Embodiment of God*[7] and *History as Apocalypse*,[8] and to do so by attempting a coherence of abstract or systematic and historical theology, although it would be better to say cultural rather than historical theology, and to emphasize that there is here an attempt to conjoin interior and theological exploration. Of course, this is saying too much, but there is no question that this is an attempt to bring Augustine and Kierkegaard together with Hegel and Nietzsche, and to do so on the ground of that Christian epic enacted by Dante, Milton, Blake, and Joyce. Such a project is uniquely my own, and perhaps it most fully occurs in this book, which might be a way of saying that my subsequent work is work in decline. It certainly would have been far wiser to have directed all of this energy simply to writing one book, but wisdom is certainly not one of my virtues, and I fear that I will leave no book that truly embodies my work. Now it is true that the last two years have been a time for me of thinking about death, and not obsessively or fearfully, but even calmly and dispassionately, but perhaps not theologically, and it could be that this is what I am now actually called to do. Is this why I sensed a summons to reread *Genesis and Apocalypse*, and is this book more actually about death than any other that I have written, and I can ask you this seriously because you know my work better than I do myself? Certainly death and nothingness are profoundly brought together in our deeper traditions, and perhaps the purest coherence between the deeply mystical and the deeply modern occurs at just this point, and only here are these ultimate worlds truly united. I must confess that I am repulsed by all that "mystical language" that has occurred in the last generation or two, a truly post-Sixties phenomenon, and I rejoice that the word "mystical' and its counterparts are alien to genuine mystical traditions.

But if I am now summoned to think about death, is not that which I have been doing all along, and doing insofar as I genuinely thought about God? Am I simply going to sing the old song once again, only this time with far less passion and power, so that I will be able to do little more than mimic or parody the theologian that I once was? This is a frightful possibility, and one that I take seriously, and so much so that it does bring forth an imperative to retire from theology. Unfortunately I am incapable of that, and deeply incapable of it, I am intrigued that even now I can only justify reading if it is somehow theological reading, so that when I read a mystery in my retirement it has to be a theological mystery. And I have little to do but to read, and to watch my precious DVDs, as I put together a collection of what I regard as genuinely theological films, and, of course, the whole world of opera and its counterparts. If all genuine festival is truly sacred, then opera is certainly sacred, and while I will surely never write about that, I could perhaps write in its spirit, and write so as to celebrate death, and celebrate it as I have not been able to celebrate it before. As you know all too well, the language of celebration has been alien to me, and perhaps most alien in *Genesis and Apocalypse*, even though the book concludes with an attempt at celebration, which can only actually speak of a No that is inseparable from a Yes, just as I have never known a redemption except that redemption which is inseparable from damnation. Indeed, if one were to speak critically about what makes my work unique, it would be its centering upon Satan and damnation, perhaps that is the point at which I am most Augustinian, and most truly a genuine theologian. Should I now move backward from this, or is the only forward movement of which I am capable a forward movement in this very direction, one that is certainly enhanced by the films that I enjoy, and it is just the absence of a sense of damnation that I find to be so devastating in our language and criticism today.

Once again I awaken to my identity as a dialectical theologian, perhaps the only one recognizing that it is Augustine's language about damnation that is his purest language about redemption, just as Kierkegaard can only actually write about grace when he is writing about *Angst* and despair, and just as Nietzsche's most purely redemptive language is his purest atheistic language, as so fully manifest in that sermon of the Madman, which has so profoundly affected us all. Could it be said that Eckhart's purest preaching is his most atheistic preaching, or that which his inquisitors could only recognize as being atheistic? McGinn is astute in drawing forth the dialectical Eckhart, but I fear that it is the dialectical Augustine who is least known in our Augustinian scholarship today, which is a major reason why it is so dull. When I was a theological student dialectical theology was the rage, then I only knew the early and not the later Barth, and only the early and not the later Luther, I didn't know Calvin at all, and didn't even after attempting to read him (surely Calvin is the dullest of all theologians). This was the spirit with which I wrote my master's thesis on Augustine's doctrine of nature and grace,[9] and if only at this point I met with nothing but encouragement from the faculty, the last time that I had a good relationship with the academic theological world. As we like to say, the demise of that world may be the most encouraging factor for theology today, and while I am the only one whom I know who truly rejoiced in his retirement from the academic world, it is possible that I cannot truly write apart from it, that I deeply need the kind of opposition that it inevitably evoked in me. So, too, it is truly a shame that the American Academy of Religion has become so dull, nothing to evoke opposition there, but could it be that I really need opposition, and deep opposition, in order to do theology?

Could it be that this is true for every genuine theology? While it is true that Aquinas is finally alien to me, I do think that his strongest writing, and strongest theological writing, occurs when he is presenting oppositions to his thinking, and perhaps this is true in one way or another of every genuine theologian. It is surely true of Augustine, who can write most passionately, and perhaps most theologically, when he is assaulting heresy, and it is his assaults on heresy that have had the deepest historical effect. Luther is only really powerful when he is assaulting the opposition; it is significant that his only real book[10] is a long and polemical assault, and an assault that in its attack upon Erasmus is an assault upon what virtually every Christian believes. Now this is a decisive clue for me to genuine theological writing, it can only be genuine if it is a profound offense, and we must remember that even Augustine's *Confessions* would have been an offense in his world, a world that had never before encountered a genuine expression of self-consciousness. So it is that it is the purely polemical Barth who is most revered, and just as this is eased in the course of his execution of the *Church Dogmatics*, so, too, does its power diminish, and only thereby can he be theologically domesticated. Perhaps what is most theological about Nietzsche is the profound offense that he creates, and just as the Nietzsche who can be taught to the average undergraduate is profoundly distant from the real Nietzsche, the theology that can be recognized by the world as theology may well be profoundly distant from genuine theology, and most so in its very language about God. I think that I would most like my theological bequest to be my assaults upon God, an assault occurring whenever I genuinely write about God, or whenever I genuinely write theologically. Could it be that this is finally true of every genuine theology, and inevitably true of any genuine theological writing, which can only be a truly and finally negative writing?

Tom

Letter 21

To Friends (June 29, 2006)

Friends,

I am reading an important book which theologians should be aware of: *The Young Heidegger: Rumor of the Hidden King* by John van Buren (Bloomington: Indiana University Press, 1994). This book is rich theologically, demonstrating how genuinely theological Heidegger's early thinking is, and how it is essentially related to the later Heidegger. Heidegger was profoundly affected by Eckhart and medieval mysticism, and more so than is commonly thought, and the metaphysics in which he grounded himself was medieval mysticism. Now we know that in 1915–1916 Heidegger's thought ultimately was philosophical mysticism, but his growing estrangement from neo-Scholasticism involved a profound confessional, theological, and philosophical conversion in the wartime years after 1916, and one issuing in his understanding of metaphysics itself as being mythical. Nevertheless, that theological and philosophical conversion is still shrouded in darkness, but Luther and Kierkegaard played a decisive role in it, and now Heidegger can call for the destruction of Christian philosophy and theology, but the end of philosophy and theology is a new beginning for it. Heidegger then embarks upon a fundamental quest for primal Christianity, and he takes the Christian experience of Parousia and Kairos and "falling" to be a specific "ontic" model for ontological categories that would make up his new beginning for phenomenological ontology.

Indeed, Heidegger intended a demythologizing of Scholasticism back to primal Christianity, and this is the origin of demythologizing or Deconstruction itself. So that Heidegger's early phenomenology of religion is at bottom a theological quest in which Angst and Death become truly central, and this issues in the birth of Heidegger as an original philosopher. Now falling, authenticity, and inauthenticity become truly primal, but primal for the first time as truly philosophical categories, eventually making possible the creation of *Being and Time*. But already in 1919 Heidegger knows an *Ereignis* entailing the abandonment of the transcendental ego, and one inseparable from *Gelassenheit*[1] and a Free Christianity and a kairological ethics, but I am getting ahead of my own reading if only to tempt you with this stunning book. Is it only philosophers who are now truly theological?

Tom

Letter 22

To Friends (July 5, 2006)

Friends,

I have continued with Van Buren, *Young Heidegger,* and am even more impressed by its theological richness and importance. Now if Heidegger is the major philosopher of the twentieth century, at least in terms of his impact, he is no less so the major philosophical theologian of late modernity, and this book identifies the creation of Heidegger's philosophy as being in large measure a theological creation (what fuel this will give the philosophical enemies of Heidegger!). For Van Buren shows in concrete detail not only the immense impact that Luther and Kierkegaard had upon the revolutionary transformation of his thinking but also the ultimate ground of that thinking in the New Testament or what Heidegger called primal Christianity, and this is true of no other modern philosopher apart from Hegel. Now it is of great significance that Heidegger focussed upon the apocalyptic Paul, here differing from his Marburg colleague Bultmann,[1] and we now can see that Heidegger is an apocalyptic thinker throughout his thinking, here, too, paralleling Hegel. This may well have occasioned his deepest break from Catholicism, but it is also fascinating that he returned to Aristotle in this period, but to a non-scholastic and demythologized Aristotle, or to a pragmatic or practical Aristotle whose metaphysics is deconstructed, and this became the original ground of what we have known as deconstruction.

Now during the period when Harnack was having an immense impact upon our understanding of the evolution of Christian dogma, Heidegger came to understand how in the very beginning Christian theology became prey to Greek "ontotheology," and a Neoplatonic God comes to dominate that theology, inevitably entailing a profound loss of the original ground of Christianity. Heidegger became determined to reverse that loss, hence the immense importance to him of Luther, Pascal, and Kierkegaard, but he transposes their theological language into a new philosophical language, and even into a new ontological phenomenology. But the creation of that phenomenology is a theological creation, or it proceeds out of the transposition of theological categories, and it is just that transposition that is the center of Heidegger's early work. And this occurs through a reversal of a theology of Glory into a theology of the Cross, entailing not only a recovery of the Crucified God, but the recovery of an original Christian faith, and one that had truly been lost or reversed in Christian tradition. Van Buren claims that the young Heidegger saw himself as a kind of philosophical Luther, dedicated to realizing an *ontological crucis*, or a radical philosophical Calvary and Good Friday, resurrected through the scandal and "foolishness" of historicity. Now a hermeneutics of death becomes overwhelmingly important, one inseparable from *Angst* and Guilt, and Heidegger sought to extract from the Christian experience of conscience the ontological condition for all ontic, moral-religious conceptions of conscience, thus making possible a kairological ethics. Accordingly, Van Buren maintains that the original awakening of Heidegger was an apocalyptic one; thence the Greek question of being was to be destroyed and repeated in terms of the Pauline-Lutheran theme of remaining awake in the night of the world for the Parousia. Being comes like a thief in the night, and Heidegger's *ontologia crucis* wanted to conceive of the mystery of *ens absconditum*[2] as enacted in and through the suffering and cross of history.

Tom

Letter 23

To Alina N. Feld* and Lissa McCullough (September 19–22, 2006)

Dear Alina and Lissa,

I am writing you as a way to address myself, and to prod myself into a new even if all-too-partial beginning, one arising from unfinished work, and from a just completed rereading of *Godhead and the Nothing*.[1] What most struck me this time was how terribly partial and fragmentary is the treatment of the advent of *Angst* in Chapter 7, "The Transfiguration of Evil." While I believe that this is genuinely original, it is so fragmentary as to have little actual meaning or effect, so it certainly must be expanded and this could give me a fresh beginning. Once again I am aware of how difficult it is to do dialectical theology today, and here is a marvelous exemplification of that, for it attempts to understand the deepest *Angst* as a consequence of the deepest Gift, and even as a necessary and inevitable consequence of that Gift, and while the impact of Kierkegaard is clear here, this is attempting to be more purely dialectical than is Kierkegaard.

The topic here is the forgiveness of sin, and the thesis offered is that *Angst* is a consequence of the actualization of that forgiveness, but it fails truly to capture that actualization, just as it also fails to be truly dialectical, thus missing or distorting the goal. But the thesis itself is important, for not only does it affirm that the origin of *Angst* is uniquely Christian, but that this origin is the birth of a full and actual Nothing, one known to Christianity alone, and one that is the necessary consequence of the actualization of the forgiveness of sin. That actualization is a consequence of Christianity's original reversal of itself, a reversal unique in world history, but a reversal that is the origin and ground of an absolute transformation of consciousness and history. So far as I know I am alone in being engaged with the question of the origin of the Nothing, or of a full and actual Nothing, or rather alone with the exception of Ray Hart and David Leahy, and I am also persuaded that the only Western philosophers who are profoundly engaged with the question of the Nothing are Hegel, Nietzsche, and Heidegger, even if this is perhaps the primal question of Eastern philosophy. Yet even as the great body of our philosophy is closed to the question of the Nothing, nothing is more primal in truly modern or late modern poetry than is the Nothing itself, a Nothing finally eliciting our most ultimate celebration, as most openly occurring in Mallarme[2] and Stevens.[3]

Now it is not insignificant that Kierkegaard and Heidegger as our deepest thinkers of *Angst* are precisely thereby our deepest modern thinkers of guilt, and a guilt that each knew as being uniquely Christian, and thereby inseparable from a uniquely Christian grace, and uniquely Christian in being dialectically inseparable from sin, a dialectic discovered by Paul and theologically created by Augustine. Hence an original forgiveness of sin is necessarily and inevitably a realization of the most ultimate guilt, one unmanifest and inactual until this forgiveness, but one now actualized to make possible an ultimate forgiveness, an ultimate forgiveness or an absolute gift only possible through the absolute transfiguration of its very opposite, and an absolute transfiguration impossible apart from the actual realization of that opposite. So it is that Christianity can know the deepest grace only by knowing the deepest guilt, a guilt that is itself a decisive sign of that grace, as so profoundly known implicitly by Augustine and explicitly by Luther and Kierkegaard. Augustine and Luther created historical revolutions by way of this realization, but this guilt can be known as an actual nothingness, this already occurs in Augustine, but it is not fully realized in thinking until the advent of German Idealism, an idealism first giving us in Schelling and Hegel an actual understanding of the Nothing, and of that Nothing which is an actual Nothing, and an actual Nothing only possible by way of an original forgiveness of sin, or an original absolute Gift.

I am intrigued by that fundamental thinking initiated by Marion on the absolute Gift, but this is a wholly non-dialectical thinking, and precisely thereby an "unrealistic" thinking, or a thinking closed to the actualities of consciousness and history. The very opposite of this occurs in Hegel, Nietzsche, and Heidegger, and if Hegel and Nietzsche are purely dialectical thinkers, Heidegger is profoundly under their impact, and thus at least implicitly dialectical. Thus in *Being and Time* the very understanding of *Angst* and death is inseparable from an ultimate call, just as in *Beiträge*[4] the understanding of *Ereignis* is the calling forth of an absolute advent that is an absolute ending, an eschatological ending that is apocalypse itself. Recent studies of the early Heidegger have fully unveiled a theological Heidegger, so that we are now coming to realize that Schelling, Hegel, Nietzsche, Heidegger, and Wittgenstein are the most theological of our modern philosophers, and also those very philosophers who have given us our deepest understanding of the Nothing. Here, the Nothing is inseparable from an absolute uprooting or an absolute disrupture, a disrupture bringing an end to every possible calm, or every possible quiescence, thus realizing the advent of actuality itself, or of a wholly negative actuality, a negative actuality that only now is all in all.

This is the actuality that we are called upon to accept as an ultimate gift, but it is a gift inseparable from death and fallenness, once again calling forth a uniquely Christian *felix culpa*, a fortunate fall that is grace itself, and a grace inseparable from an ultimate *Angst*. Now if it is the Incarnation that Christian thinkers commonly know as the absolute paradox, in this perspective that paradox can be known as a universal paradox, a paradox manifest in every actual opening to either sin or grace, and every actual opening to actuality itself. Already Paul deeply knows this, and knows it in knowing the advent of an absolute grace that is inseparable from the advent of an absolute guilt, so if it is Paul who is the source of what Nietzsche so profoundly knows as the "bad conscience," he is even thereby the original unveiler of what we have most deeply known as grace, an unveiling inseparable from the unveiling of an absolute guilt. For the first time many of our most creative philosophers are now being profoundly challenged by Paul, and if Nietzsche is our most deeply Pauline late modern thinker, he can be so only by way of the mirage of being a pure enemy of Paul, an enemy releasing *The Antichrist,* which many now regard as his greatest work. Of course, it was Paul who made possible the Augustinian revolution, a revolution inaugurating a uniquely Western thinking, and a uniquely Western dialectical thinking, so that a Western dialectical thinking can be understood as a genuinely Pauline thinking, and this is just as true of Hegel as it is of Kierkegaard and Nietzsche.

Now if a fully actual nothingness only becomes historically incarnate with the advent of Christianity, this advent is inseparable from the forgiveness of sin, and the apocalyptic forgiveness of sin, a forgiveness occurring in the depths of "darkness," and one therein calling forth those very depths, depths invisible and unheard apart from such an apocalyptic forgiveness, but depths now speakable and hearable as darkness itself, or as a full and actual nothingness. Just as Hegel in the conclusion of the *Science of Logic* could understand the "cunning of reason" as a reversal of every actual intention and goal, the advent of an actual nothingness can be understood as a reversal of the apparent goal of forgiveness, but it is a reversal necessary for an absolute or apocalyptic forgiveness, and necessary if only to actualize that absolute guilt or that absolute darkness that is negated in an absolute forgiveness. But what is that negation? It certainly cannot be a dissolution of that guilt or darkness, and cannot be if only because the reverberations of that negation are historically so overwhelming, reverberations inaugurating an absolute guilt and a fully actual nothingness, thus wholly transforming that consciousness and that history that are here at hand.

Thus an apocalyptic forgiveness of sin cannot be understood as a dissolution of sin but far rather as the transfiguration of sin, a transfiguration realizing an absolute reversal of sin, a reversal in which the depths of sin are reversed into the depths of grace, thus creating an absolutely new grace, and a new grace that is only manifest and actual as such upon the horizon of a total darkness. Accordingly, the New Testament is unique in scripture in calling forth a total guilt and a total darkness, at no other point does it so clearly differ from the Old Testament or from world scripture, but such a darkness is a necessary horizon for a witness to or calling forth of a total forgiveness or a total grace, and if that grace is apocalypse itself, a genuine apocalypse is absolute light and absolute darkness at once and simultaneously. This is a simultaneity that can be discovered not only in the epic enactments of Dante, Milton, Blake, and Joyce, but in Nietzsche's enactment of

Eternal Recurrence and Heidegger's celebration of *Ereignis*, all of which can be understood as consequences of the advent of an actual nothingness, or of an absolute transfiguration realized only in and by that very advent. Hence an *Angst* arising from an encounter with the Nothing is only made possible by an apocalyptic forgiveness realizing the horizon of an actual nothingness, an horizon apart from which the Nothing would be wholly inactual and unreal, but within which the Nothing is inevitably manifest, or manifest to those who are awake.

Luther, Kierkegaard, and Heidegger can all know *Angst* as effecting the deepest possible call or summons to us, this is a call only possible as a consequence of the advent of a full and actual Nothing, a nothingness actualized by the forgiveness of sin, and by that forgiveness of sin that is an absolute transfiguration. Only now does darkness itself become incarnate, which can be understood as a consequence of that Incarnation which culminates in the Descent into Hell, a Descent into Hell that is a Harrowing of Hell, a harrowing that is not only virtually absent in Christian theology, but in Christian imagery and iconography as well. Why such a profound silence? Indeed, the New Testament itself is almost silent on this crucial front, and this despite the fact that the New Testament is more centered upon damnation and Hell than any other scripture in the world. Nevertheless, Paul could deeply affirm, and do so in the very context of his purest apocalyptic writing, that for our sake God made Christ who knew no sin "to be sin" (2 Cor. 5:21), but this statement is unique in the New Testament and has apparently never called forth an actual Christian imagery. Yet such an affirmation is clearly a dialectical one, and if Paul is the creator of a uniquely Christian dialectic, and is so by inaugurating the apprehension of a true dialectical movement in which opposites actually pass into each other, so that the depths of sin pass into the depths of grace, even as the depths of grace pass into the depths of sin, then we can understand not only that the opposites of sin and grace are actually inseparable, but that their very full and final actualization embodies an ultimate *coincidentia oppositorum*.

Here, too, lies a uniquely Western and Christian *coincidentia oppositorum*, realizing interior depths that are truly dichotomous depths, wherein an ultimate transfiguration can occur, and a transfiguration of the ultimate depths of evil itself. No such transfiguration can occur apart from the realization of the depths of nothingness, therein making possible the advent of what a deeper imagination and a deeper thinking can know as the Nothing, yet a Nothing that is the very opposite of nothingness, or the opposite of a pure nothingness, and is so by way of its own ultimate actuality. A Christian naming of the Nothing occurs in a Christian naming of Satan, but at no point is our theology weaker and more fragmentary than it is in its understanding of Satan, nor at any other point does it stand so wholly removed from our imaginative creations, imaginative creations that in full and late modernity call forth ultimate enactments of Satan and Hell. But even as this occurs theology ever more fully removes itself from any possible understanding of Satan and Hell, even muting itself on the subject of damnation, which is truly central in all premodern theology, thus giving us a theology that is wholly closed to an absolute transfiguration, and is so closed if only because it is alienated from the ultimate depths of evil and darkness.

Of course, there is little understanding of evil in our philosophical traditions, an in depth philosophical understanding of evil not occurring until German Idealism, and then disappearing in twentieth century philosophy, and this despite the profound and overwhelming enactments of evil in the late modern imagination. Now we cannot say that it is impossible to think the depths of evil if only because this so fully occurs in Schelling, Hegel, and Nietzsche, and if these are the very thinkers who embody the depths of the imagination, that is a depth inseparable from the depths of evil, depths of evil making possible that absolute transfiguration occurring throughout our greatest imaginative creations. It is a philosophical understanding of that transfiguration that makes possible German idealism, as most purely and most comprehensively occurring in Hegel's dialectical philosophy of absolute negation, an absolute negation that is an absolute transfiguration, and an absolute transfiguration of absolute opposites. While it is Nietzsche who most ultimately inverts and reverses this dialectical negation, and this despite having virtually no acquaintance with Hegel, this is an absolute inversion that is nonetheless an absolute negation, and a dialectical absolute negation, as an absolute No-saying passes into an absolute Yes-saying, a Yes-saying wholly impossible apart from that No-saying.

Nietzsche's enactment of Eternal Recurrence is an enactment of absolute transfiguration, a transfiguration impossible apart from a calling forth of the depths of abyss and darkness, depths that are themselves transfigured in this enactment, as here the deepest darkness passes into the deepest light, or the most absolute No is transfigured into an absolute Yes. While Nietzsche apparently stands alone as the philosopher of Eternal Recurrence, the truth is that Eternal Recurrence fully parallels a uniquely modern dialectical negation, so that Hegel and Nietzsche can be understood as dialectical twins, and if each is the very opposite of the other, these are opposites that nonetheless coincide, and do so as purely dialectical thinkers. Only Heidegger has fully conjoined these thinkers, and done so as a thinker who is simultaneously ontological and historical at once, and simultaneously systematic and anti-systematic at once, thereby conjoining our greatest modern systematic thinker and the greatest anti-systematic thinker who ever lived. So, too, Heidegger is a deeply theological and deeply anti-theological thinker at once, thereby once again conjoining Hegel and Nietzsche, and if all three of these thinkers profoundly know the death of God, each of them can finally realize that death as an ultimate liberation.

This is the liberation that dawns as the forgiveness of sin, and the apocalyptic forgiveness of sin, and if that forgiveness is inseparable from an enactment of the deepest depths of darkness, depths called forth by that very forgiveness, a philosophy of absolute transfiguration transfigures these very depths, and this occurs in the deeper thinking of Hegel and Nietzsche and Heidegger alike. So, too, does it occur in our deepest imaginative vision, one realized philosophically by Schelling and Hegel, and by Nietzsche and Heidegger, too, and these are our only philosophers who embody imaginative vision. Yet transfiguration is ultimately and at bottom a theological rather than philosophical category, it does not occur in our philosophical dictionaries, perhaps because it is inseparable from a theological or ultimate redemption, and a redemption from the depths of evil and abyss.

Dialectical philosophy has itself perished in the twentieth century, or perished insofar as it is not embodied in Marxism, but Marxism has given us our deepest and most comprehensive understanding of evil in the twentieth century, and if Marxist enactments of redemption are truly and purely reversed in that century, nothing else has so darkened the very idea or symbol of redemption.

Now not only does Nietzsche fully and ecstatically employ a language of redemption, but this at least partially occurs in Heidegger, too, and explicitly so whenever he speaks of *Ereignis*, and this occurs far more decisively in these thinkers than it occurs in our modern theologians, or theologians apart from that solitary genius Karl Barth. Perhaps no language has been so emptied in modernity as has theological language, an inevitable consequence of a uniquely modern realization of the death of God, so that its occurrence in our philosophical thinkers is far more powerful than it is in virtually all of our theology, but the question must inevitably be asked if this is a genuine language of redemption. Here, Hegel has inevitably drawn the deepest assault, but that assault should necessarily be directed at Nietzsche, too, and here remarkably there is a deep resistance to such assault, and perhaps it is impossible to deeply assault Nietzsche, or impossible either to annul or to dissolve his language. If so, this bespeaks a truly miraculous power, one obviously absent in Schelling and Hegel, and it gives us an insight as to why Heidegger was impelled to create the densest and most difficult of all philosophical languages.

The late Nietzsche was much given to unveiling that absolute reversal that Christianity has effected upon Jesus, thereby Nietzsche comes to know a Jesus whose praxis is wholly centered upon forgiveness, a forgiveness that he silently or indirectly knows as the initial enactment of the death of God, or the death of the absolute Creator and Judge. Indeed, in *The Antichrist* (§33) Nietzsche can proclaim that in true Christianity, one only practiced by Jesus, sin is abolished, and precisely this is the "glad tidings," for blessedness is not promised, it is the only reality. This is the blessedness discovered once again by a new Zarathustra, a new Zarathustra who enacts blessedness as Eternal Recurrence, and thus a blessedness whose only portal is through an absolute horror, and the absolute horror of actuality itself. This is the horror that is ultimately blessed by an enactment of Eternal Recurrence, and just as no one has known a *horror religiosus* more deeply than did Nietzsche, this is a horror that first comes to an end through the forgiveness of Jesus, even if it is only in Paul that this horror is first purely spoken. If Nietzsche and Blake are commonly known as our greatest modern prophets, at no point are they so fully united as in their ultimate enactment of an absolute forgiveness, a forgiveness only occurring through what Nietzsche knows as the death of God and Blake envisions as the "Self-Annihilation" of God, and a death releasing what Nietzsche knows as an ultimate nihilism and Blake envisions as a total embodiment of Satan.

But that embodiment of Satan is so only as a "Self-Annihilation" of Satan, just as that ultimate nihilism is one reversed in Eternal Recurrence, and a reversal only possible through that nihilism itself, just as the "Self-Annihilation" of Satan is possible only through the total embodiment of Satan. For Blake and Nietzsche are those prophets who most ultimately envision and enact an absolute darkness, an enactment inseparable from either Eternal Recurrence or Self-Annihilation, and thus inseparable from everything that here can be known as an absolute forgiveness. It is remarkable how primal Jesus is both to the late Nietzsche and to the mature Blake, and both most deeply associate the name of Jesus with an absolute forgiveness, and for both an absolute forgiveness wholly reversed by both Christianity and our history itself. So, too, Blake and Nietzsche are our most offensive prophets, enacting what our common experience and our common sense can know as pure blasphemies and pure assaults, and pure assaults upon our most deeply cherished assurances and beliefs. Even if biblical scholarship can ignore each of these prophets, biblical scholarship has never given us a deep understanding of prophecy, and when one does occur in Max Weber, it does so under the impact of Nietzsche.

Is it possible today to speak of any kind of ultimate advent or any kind of ultimate beginning? Perhaps at no other point are we so distant from Hegel and Nietzsche, or so distant from a genuine historical consciousness, or from that genealogy which is so ultimately important for Nietzsche. Indeed, Nietzsche created everything that we more deeply know as genealogy, a genealogy that is a dialectical genealogy, and dialectical in knowing a genuine or ultimate origin as being the very opposite of all apparent or manifest origin. Hence it is quite Nietzschean to know the advent of *Angst* as being the consequence of an absolute forgiveness, or to know the origin of an absolute forgiveness itself as being inseparable from an ultimate curse. No one has known curse itself so purely as did Nietzsche, a curse that he could know as a universal curse, one occurring throughout all history and consciousness, and one whose origin is an original fall. It is truly remarkable that Nietzsche has known an original fall more deeply than any theologian, even including Augustine, and it is that fall that he enacts in creating an ultimate genealogy, one that is nowhere thought so deeply as it is in *The Genealogy of Morals*.

Augustine himself can appear as an innocent in this perspective, a perspective profoundly reversing everything that we have known as ethics, and doing so precisely in calling forth the origin of that ethics. Here, it becomes fully understandable why we should so deeply resist the question of origin, a question only now becoming an absolutely forbidden question, and forbidden because when origin becomes deeply investigated it calls forth a true horror or a pure assault. Are our most deeply cherished ideals the very opposite of their manifest meaning, and do we inevitably come to know this when we realize a genuine genealogy, a genuine genealogy not created until Nietzsche, but thereafter it has had an overwhelming effect, and not least so by psychoanalysis? Nowhere is such a pure *dysangel* or such an overwhelmingly "bad news" more deeply embodied than in Nietzsche, and yet no thinker has been a more liberating thinker than Nietzsche, and most liberating to those most overwhelmed by *Angst*. The very category of fall, and of ultimate fall, opens up a new perspective upon *Angst*, for if *Angst* is a consequence of fall, then in a Christian context it can be known as a *felix culpa* or fortunate fall, a fall realizing that very consciousness that is transfigured in an absolute forgiveness or an absolute redemption, and a redemption and forgiveness impossible apart from ultimate fall.

In the deeper expressions of the Western imagination, and the deeper expressions of Western thinking, too, an enactment of ultimate transfiguration is inseparable from an enactment of ultimate fall, this is perhaps clearest in Hegel, Marx, and Nietzsche, but most overwhelming in Dante, Milton, Blake, and Joyce, and yet it is Nietzsche who most decisively and most comprehensively thinks the fall, a thinking apart from which a total Yes-saying would be simply impossible. Hence any liberation that is here promised is inseparable from the most ultimate fall, only an awakening to the depths of fall makes true liberation possible, so that this awakening is essential to a genuine liberation. So likewise it is here essential to know ultimate fall as an original fall, a fall only possible as an original fall, for only if it is an original fall can it actually be reversed, for liberation is possible only from an original and not from an eternal fall. So, too, it is only an original and not an eternal fall that we ourselves can enact, only if it is an original fall is it possible for us to be responsible for fall, a responsibility that we know in knowing an ultimate guilt, and apart from that guilt there can be no ultimate transfiguration.

There is a deep Western tradition centering upon absolute guilt, one initiated by Paul, and then profoundly carried forward by Augustine, Luther, Kierkegaard, Nietzsche, and Heidegger, but this is simultaneously a tradition centering upon an ultimate liberation, and not only a liberation from that guilt but a liberation transfiguring that guilt. This transfiguration is possible only if guilt is an actual guilt, and an actual guilt that is a truly actual nothingness, a nothingness that is not a privation but a full actuality, hence one inseparable from the Nothing itself. This is a Nothing that dawns ever more decisively in this tradition, and if it is only fully thought in Schelling, Hegel, and Kierkegaard, it is decisively enacted in Nietzsche and Heidegger, an enactment inseparable from everything that here can be realized as liberation. Thus the Nothing is truly essential to an ultimate liberation, hence its centrality in our deepest mystical enactments, and just as Eckhart is truly primal for Schelling, Hegel, and Heidegger, he is so most deeply in knowing Godhead itself as the Nothing, but this is an absolutely liberating nothingness, and therefore an absolutely actual nothingness. Indeed, to apprehend the Nothing, even if doing so only in the deepest *Angst*, is to know a power that can shatter our deepest roots, and therefore shatter our deepest bondage.

Now such shattering could not be a simple dissolution, could not be a simple or literal ending or death, and could not be because of the very power of an actual nothingness, a nothingness reflecting an absolute nothingness, so therefore its realization could never result in a simple dissolution, and could never do so if only because of the ultimate power of nothingness itself. While it is extraordinarily difficult genuinely to know this power, although it is purely known by Schelling and Hegel, this is a power that is inevitably apprehended in knowing or realizing *Angst*, and *Angst* is universal in a truly fallen world, or in a world that is the consequence of an original fall. If *Angst* is a consequence of an encounter with the Nothing, it is certainly a profound witness to an original fall, for the very power of *Angst* is inseparable from an ultimate or absolute guilt, a guilt that could only be the consequence of an original fall, and a guilt that we know as our own in deeply knowing *Angst*. Hence those thinkers who have most deeply understood *Angst*, Kierkegaard and Heidegger, have called forth the deepest understanding of responsibility itself, and done so in their very understanding of guilt. This is not only a genuine renewal of Augustine and Paul, but even thereby a genuine renewal of a uniquely Christian grace, and the uniquely Christian grace of forgiveness, a forgiveness only actualized through the depths of guilt, and a guilt whereby we can deeply know that which is most genuinely our own.

It is not insignificant that it is in our great tragedies, and above all in our great modern tragedies, that guilt is most overwhelming, and most overwhelming as an interior guilt calling forth the deepest depths of the individual, an individual who is most uniquely individual in those very depths, depths that unveil an individual more decisively than anything else, so that it is not until the realization of this ultimate guilt that a uniquely individual consciousness first dawns. Augustine's *Confessions* records this dawning, but it is Shakespearean tragedy that first profoundly embodies the truly unique individual, and does so through a tragic action revolving about an ultimate guilt that is an ultimate responsibility, and a responsibility that can only enact itself as a tragic responsibility. But that is a tragedy releasing an ultimate assuagement, and an assuagement only possible through a tragic destruction, a destruction that is perhaps the purest witness to the Nothing, and yet a witness unveiling the overwhelming power of transfiguration, and a transfiguration only possible through the Nothing.

Nothing has so resisted understanding as has tragedy itself, perhaps because we so resist understanding the Nothing, but an evocation of the Nothing does call forth a way both into Greek and into modern tragedy, and if criticism has already established Sophocles's *Oedipus at Colonus*[5] as a celebration of the Nothing, we surely could understand Shakespeare's *King Lear* as such a celebration, and thereby become open to tragedy as an ultimate reconciliation that is an ultimate transfiguration. Certainly tragedy calls forth the ultimate power of an absolute negativity, and if that negativity ultimately derives from the Nothing, or from an actual realization of the Nothing, then the positive power of the Nothing thereby becomes manifest, and manifest in that ultimate assuagement that is unique to tragedy. Can we understand tragic assuagement as an enactment of an absolute forgiveness, or of an apocalyptic forgiveness, a forgiveness that is an apocalyptic *coincidentia oppositorum*, a coincidence of absolute destruction and absolute fulfillment, wherein each is wholly unreal apart from the other? If we can understand that tragedy evolved out of primordial sacrifice, and that tragic action is at bottom a sacrificial action, and an action realizing a total presence, is that the total presence of an absolute transfiguration only made possible by the Nothing itself?

Tom

Letter 24

To John B. Cobb, Jr.* (September 27, 2006)

Dear John,

I am now rereading Ray Monk's biography of Wittgenstein[1] and writing to urge you to read it if you have not yet done so. Frankly I was never able genuinely to respond to Wittgenstein's writing even if I became aware that there are deeply religious motifs within it. But Monk's book, which has been critically hailed as the best of all biographies of a philosopher, overwhelms one with the details of Wittgenstein's quest for perfection, and a perfection that is ethical and religious at once, and one that was even the ultimate goal of his writing. He himself would publish very little, despite the enormous and worldwide interest in his work, but he wrote a vast body in his diary and journals, and these form the primary data for Monk's book, *Wittgenstein: The Duty of Genius*. I trust that you are aware of how much he loathed the high academic world, and that he broke with Bertrand Russell because Russell was not a Christian (and hence could not possibly be a serious thinker). And while Wittgenstein avoided church, and all open Christian fellowship, his private writings reveal him as a profoundly committed Christian, and he continually attempted to realize that ground in philosophical writing itself.

Not only is his purely negative thinking a genuinely iconoclastic thinking, but it is intended to release a pure thinking that while not about ethics or religion, nonetheless calls ethics and religion forth, as a genuine religion would immediately be actual if our thinking and our consciousness did not turn us away from it. You will remember that while young W renounced a vast fortune and ever after most frequently lived a life of poverty, he also continually tried and failed to live with and as common or poor people, and thought that at most he could only be a worker or a grammar school teacher. While Trinity College at Cambridge where he was given a doctorate and then a Fellowship was then a peak of the academic world, it became a Hell for him, and most of all because of its hypocrisy. W was largely treated as a god by his students and peers but he found that virtually none of them could understand his thinking, and he continually ravished himself by way of his compulsive quest to realize a genuinely philosophical writing, one that he was persuaded had not yet occurred. You know, of course, of how deeply he turned away from his early thinking, but this was a gradual and a torturous process, and is never fully achieved.

Wittgenstein also profoundly turned against the modern world itself, he quite early realized that the whole modern conception of the world is an illusion, and its so-called laws of nature are not the explanation of natural phenomena. For the meaning of the world does not lie within it but outside it, and that meaning we can call "God," so that to genuinely think about the meaning of the world is to pray, and whereas I am completely powerless, I can make myself independent of the world by renouncing any influence upon its happenings (pp. 140ff.). These reflections occurred when he was an Austrian soldier during the First World War, and could know conscience as the voice of God, just as the good life, which Wittgenstein ever profoundly sought, is the world seen *sub specie aeternitatis.*[2] Now Wittgenstein in his solitude was deeply devout, much given to reading or meditating upon the Bible, and certainly deeply in quest of salvation, and even salvation through Christ's Resurrection. But this quest is inseparable from a continual and torturous realization of Angst and guilt, which is paralleled by a continual realization of the illusions of our thinking, he is almost a Buddhist in that realization, but this is a dialectical thinking, for the genuine realization of illusion is the realization of a genuine thinking that is a dissolution of illusion, and that thinking can be known as the mystical itself.

Unfortunately there is no book of his that decisively records this quest; *Philosophical Investigations* is a hodgepodge of various excerpts from the journals and only its first part has any real coherence, and Wittgenstein could never accept another's interpretation of his work, even though many powerful minds gave themselves to it. One thing is clear however, here genuine thinking is inseparable from genuine praxis, indeed, genuine thinking *is* praxis, and is a liberating praxis, and we are overwhelmingly in need of a total liberation. From one point of view Wittgenstein was the most pathological of all thinkers, even including Nietzsche, and genuinely possessed as was no other major thinker, but that obsession is finally an obsession with "God," and with the uniquely Christian God, a God whom he can recognize as being truly portrayed in *The Antichrist,* and a God truly called forth by Tolstoy and Dostoyevsky, so that Russia became a sacred goal for him, and he even visited Soviet Russia where he wanted to be a worker in a futile attempt to realize for himself a sacred Russia.

Now it is clear that this is not the Wittgenstein who is known in analytic philosophy, perhaps that philosophy is further from his than any other, and yet it is also clear that this is ultimately a profoundly solitary thinking, even if it is a thinking that could become real for any of us. Let us also face the fact that Wittgenstein is at least in part a theologian, and one truly unrecognized as such.

Tom

Letter 25

To John B. Cobb, Jr. (October 4, 2006)

Dear John,

I have just finished rereading *Religion in the Making*,[1] or reading it for the first time, since I was in my Neo-orthodox phase when I first read it and thus closed to its real thinking. I am startled that there is so much about God here, a new thinking about God, and hence inevitably strange. But what most strikes me now is its substantial sections on evil, I have long thought that only Schelling, Hegel, and Nietzsche among philosophers are capable of thinking about evil, and it appears odd to place Whitehead in this company even if I have long thought that Whitehead should be thought of along with Hegel. Surely Whitehead is alone among twentieth-century philosophers in genuinely thinking about evil, but here at least it is clear that this is inseparable from his thinking about God. And oddly enough I have a truly new sense of empathy with Whitehead here, for I think that he is most concerned to think about the transmutation of evil, and the transmutation of evil into good, which occurs in actuality through God, a God who not only has in his nature the knowledge of evil, but evil itself becomes a stepping stone in the realization of God (although here I may be misinterpreting p. 149).[2]

Now I thought that I have been virtually alone as a theologian in coming to center upon the transfiguration of evil, a transfiguration occurring in Godhead itself (*Godhead and the Nothing*[3]), but it would appear that I have a predecessor in Whitehead. Or is this untrue? And is his thinking of evil here truly inseparable from his thinking about God? I note once again that Whitehead has a most acute sense of how offensive Christianity and the Christian God have become in our world, and while I sense that he was unacquainted with that modern literature that draws forth the *horror religiosus* of Christianity, he nevertheless knew all too deeply that Christian God who cannot actually be known in modernity, and I sense that a decisive dimension of his thinking about God occurs in a dissolution of that God. Surely only such a dissolution could make possible a thinking of the transmutation of evil in God, just as only such a thinking could call forth the actuality of evil, one truly closed to our theologies as I attempt to demonstrate in *Godhead and the Nothing*. But it occurs to me that there is an apologetic possibility here that Whiteheadeans have apparently ignored, for if this "idealist" is open to the actuality of evil as is no other late modern philosopher, then this is a uniquely contemporary "realism," and realistic as is no other philosophy upon our horizon. Now I know that you have been maintaining this for years but for me at least this truly becomes actual and concrete at this point!

I note that it is in his treatment of evil that Whitehead can most to decisively know Christianity as a doctrine of escape, yet he knows Christianity as a religion seeking a metaphysic, whereas Buddhism is a metaphysic generating a religion. And isn't it marvelous that he spends so much time correlating Christianity and Buddhism, and understands that both have decayed in modernity, and done so in part through ignoring each other. But I must ask you about his treatment of evil (p. 92), when he speaks here of a complete determinism as meaning the complete self-consistency of the temporal world, a conclusion of all thinkers who trust metaphysical concepts, he goes on to say that the difficulty of this conclusion comes when we confront the theory with facts, and by reason of the necessity of confirmation with the nature of God, for then the evil of the world is in conformity with the nature of God.[4] Now I know that Whitehead is not a "determinist," but isn't evil in conformity with the nature of God, and must be if only because of God's transmutation of evil, a transmutation apart from which we finally cannot understand evil at all?

Yours hopefully,
Tom

Letter 26

To Andrew W. Hass*
(January 5, 2007)

Dear Andrew,

You know that I genuinely respect your theological integrity and power, but now I am addressing it in an attempt to renew my own theological thinking, one that I have feared has come to an end, and perhaps it is an illusion to think otherwise. I wish now to carry forward my all too recent thinking about Hegel, and specifically to deal with the question or problem of an actual nothingness, and even an absolutely actual nothingness. So far as I know I am the only Hegel interpreter who has engaged this question, and I advanced the thesis in *The Genesis of God* that it was Hegel in the *Science of Logic* who philosophically discovered an actual nothingness, a nothingness apart from which a genuine or ultimate self-emptying would be impossible. Now I have also attempted all too partially to deal with Augustine on this crucial point, and while it would appear that his Neoplatonism makes a thinking of an actual nothingness impossible, the depth of his understanding of sin, a truly new depth, inevitably evokes the horizon of an actual nothingness, a horizon that is first imaginatively envisioned in the *Inferno*, a truly revolutionary breakthrough to an ultimate nothingness, and one giving birth to a uniquely Western *mimesis*. Thus all too paradoxically our uniquely Western realism is born by way of an epic enactment of Hell, or an epic enactment of an actual and even total nothingness, a nothingness that is the very arena of a uniquely Western actuality.

As my theological work has evolved, it became ever more fully centered upon that actuality, one impossible to understand theologically apart from the death of God, and it is not insignificant that Hegel as the first philosopher of the death of God is the first philosopher of a uniquely Western actuality, one that can only be understood as such historically, so that Hegel was inevitably the creator of the philosophy of history. While I certainly have not read most of the vast body of writing on Hegel, never in what I have read have I encountered a truly serious study of his philosophy of history, and I am tempted by the idea that here history cannot actually be understood apart from an actual nothingness, an actual nothingness that historically was inaugurated by Christianity, as so profoundly understood by Nietzsche, and it is that very Augustine who is the first thinker to become open to an actual nothingness who created the theology of history in the *City of God*. As you know, I understand both Hegel and Nietzsche as being profoundly Augustinian, and both are so most explicitly in centering upon self-consciousness, a self-consciousness that each in their all too unique ways could understand as being inaugurated by Christianity, but what few have understood is that an Hegelian self-negation or self-emptying is impossible apart from the horizon of self-consciousness, which is a fundamental reason why it has become so vacuous in a world in which self-consciousness is ending. While all serious Hegelians are aware of the integral relationship between the *Phenomenology of Spirit* and the *Science of Logic*, this relationship has only recently begun to be critically explicated, and one could expect that there will occur a closing of any possible gap between "self-negation" and "self-emptying," a closing that could draw forth a truly new realization of self-consciousness itself.

Now a fundamental difference between Hegel and Nietzsche is that Hegel is absolutely affirmative of self-consciousness, knows its full actuality only to be born with the advent of Christianity, an advent that is the ultimate center of history itself, whereas Nietzsche is absolutely negative towards self-consciousness, knowing it to be born with the advent of *ressentiment*, thereby knowing it as the ultimate origin of our deepest imprisonment, and thereby knowing the depths of our will itself as an absolutely destructive and self-lacerating will. Yes, we want to think that will is an illusion, but that is a way of closure to the deepest horror within us, and a horror from which we can be released only by a total affirmation of the death of God, and the death of that God who is the origin of will itself. If Nietzsche is the last philosopher of the will, he thereby is the successor to the first philosopher of the will, Augustine, and if it is Augustine and Nietzsche who have given us our deepest understanding of an ultimate self-laceration, it is Nietzsche and Augustine who are our purest thinkers of that unique grace that is a total reversal of self-laceration, and thereby a reversal of everything that we have actually known as either selfhood or self-consciousness. At no other point has Nietzsche's impact so transcended Hegel's in late modernity, just as it is precisely here that a Nietzschean thinking has so fully been embodied in the late modern imagination, and if that imagination finally enacts a total vanishing of self-consciousness, that is a vanishing making possible the deepest affirmation, or an absolute Yes only made possible by a total enactment of an absolute No.

Indeed, it is only in the deepest expressions of the late modern imagination that an actual and total nothingness is fully realized or enacted, and simply to name our greatest artists and visionaries is to name our deepest enactors of such a nothingness, and yet we can know that this enactment is a consummation of an imaginative tradition inaugurated by Dante himself, or by Dante and Giotto.[2] But Giotto and Dante are inaugurators of imaginative realizations of self-consciousness, as for the first time self-consciousness is decisively realized in painting and poetry, and just as this realization evolves ever more fully and finally in the Western imagination, it is finally consummated in late modernity, and consummated in imaginative works defying virtually all theological acceptance or understanding. Yet it is possible to know our greatest late modern imaginative realizations as enactments of the death of God, this often occurs in many of the finest critical studies of these works, but thereby the death of God, or a uniquely modern realization of the death of God, can be known as the origin of an absolutely new and total nothingness, one ushering in a new and total nihilism, a nihilism simply inescapable for us, and one that Nietzsche so profoundly and so prophetically already understood. Now even if such nihilism is wholly alien to Hegel, Hegel's "absolute idealism" can be understood as an anticipatory reversal of such nihilism, and just as this is an "idealism" embodying an actual and total nothingness, a nothingness making possible its ultimate self-negation or self-emptying, this is an "idealism" occurring in the very arena of nothingness itself. We can even understand the *Commedia* as the origin of modern idealism, but a *Commedia* inaugurating *mimesis* itself, and thus inaugurating a uniquely Western realism, a realism not philosophically understood or enacted until Hegel, thus making possible our most ultimate modern philosophical revolution.

Unfortunately this philosophical revolution has not been accompanied by a theological revolution, and in its wake theological thinking has become ever more distant from the actualities of our world, and above all distant from a uniquely modern nothingness, even if we now can know that an absolute nothingness is at the very center of our deepest and purest mystical ways. It is intriguing that Hegel could be far more open to a deeper mysticism than can our theologies, or our manifest theologies, and this could be said of Nietzsche, too, perhaps it is depth itself that is most resisted by our theologies, and above all so at a time in which our manifest theologies are only conservative or reactionary theologies, or can be genuine theologies only by dissolving all language about God. But it is significant that our seemingly most influential contemporary theology, radical orthodoxy, came into existence by way of a deep revulsion against modern nihilism, and can even know modernity itself as an embodiment of nihilism, so that its quest for postmodernity, even as is true of other conservative theologies, is a quest for the dissolution of modernity. Of course, such a quest is the very center of fundamentalisms throughout the world, and the very power of these movements is a decisive witness to the ultimacy of a uniquely modern or uniquely contemporary nothingness, and a nothingness that is not only a theological nothingness, but a political nothingness as well, as so forcefully embodied in a uniquely modern totalitarianism, and a totalitarianism whose echoes so decisively challenge us today.

Yours faithfully,
Tom

Letter 27

To Andrew W. Hass (January 11, 2007)

Dear Andrew,

Now that I have been mulling over your extraordinary Hegel letter for several days it is time to respond. First, as I remarked, at the present time I am again overwhelmed with the challenge of nihilism leading me emphasize far too much the role of an actual nothingness in Hegel. I certainly do not see it as the hub of his system, but I continue to believe that Hegel's philosophical discovery of an actual nothingness, one making possible a profoundly revolutionary ground of his logic, is ultimately important. So, too, I do not think that we can decisively understand a uniquely Hegelian negativity, or actuality, or opposition, apart from its ground in a fully actual nothingness. Yes, I think that you are right in asserting the profound importance of Aristotle for Hegel, he can even be understood as our greatest modern Aristotelian thinker, and our greatest anti-Platonic or anti-Neoplatonic thinker. But I resist associating Hegel with Plotinus, in part because I have never been able really to enter Plotinus, and also because Hegel is so deeply anti-Neoplatonic. This is an absolutely fundamental point theologically, one apart from which we cannot become open to his realization of the death of God, or his ultimate negation of absolute transcendence.

Yet I continue to think that an absolutely fundamental question in Hegelian thinking is the question of genesis or origin, one that is paramount both in the *Phenomenology* and the *Science of Logic*, and that this ultimately opens the question of the origin of actuality, of negativity, and of opposition. However, I cannot resist thinking that here Hegel deeply censors or disguises his own thinking and writing, for not only does this impel the question of the origin of God, but the origin of Godhead itself, and not simply the question of the origin of abstract Spirit or the Bad Infinite. Note how in those tantalizingly few pages that conclude the discussion of revealed religion in the *Phenomenology of Spirit*, he does forcefully speak of the alienation of the divine Being, and a self-alienation yielding to death, a death that is a reconciliation of absolute essence with itself, and this was initiated by a divine Being that *from the beginning* externalized itself, becoming self-centered and "evil." As you know this section has been extraordinarily important for me, but apparently for no other theologian. But then there is a common resistance to entering the theological Hegel, or even to acknowledging that Hegel was a profoundly theological thinker, and a profoundly original theological thinker, perhaps the most original theological thinker who ever lived.

Now I much admire your goal of bringing art and religion together in Hegelian thinking, or in a "surpassed" Hegelian thinking, and there surely is a profound weakness here in our understanding of Hegel, which Agamben[1] cannot address because he is closed to theological thinking even if he does have a deep understanding of nihilism. I think that a major problem in the interpretation of Hegel is that subjects such as art and religion are investigated almost solely by way of Hegel's university lectures, largely if not wholly ignoring his major works, but the truth is that literature and perhaps art exercise a profound role in the *Phenomenology*. I like to say that this is the first major philosophical work not only to become open to but actually to enact the imagination, and this is one of its most important revolutionary grounds. Note the gulf that here lies between the revolutionary Hegel and the Kant of the third critique, even if that critique does initiate a full philosophical aesthetics, and here I think that there is a world of difference between Kant's historical world and Hegel's, for Hegel's fundamental thinking was born by way of the impact of the French Revolution. As you know I don't think that it is possible to understand Hegel apart from the French Revolution, and that he remains our most profound interpreter of that revolution, an interpretation that made Marxism possible, even if few Marxists and above all the followers of Althusser[2] can accept this.

I must confess that it has been many years since I read Hegel's lectures on aesthetics—I lost my copy in the fire[3]—but I know that I was once deeply affected by them, and most so by his understanding of Greek art and literature. The section on the Classical world is the great weakness of Hegel's lectures on the philosophy of religion, and this despite the fact that Hegel was an important Greek scholar, here perhaps his deep Christianity blinded him, but this is not true in the lectures on aesthetics. Of course, this was a period when the German intellectual world virtually worshipped Greece, but I wonder if you have been affected by his understanding of Greek comedy in the *Phenomenology*, which is perhaps his purest enactment of the death of God, and surely the most profound understanding of comedy that we have been given. Note how here art and religion are so integrally conjoined, and also in his understanding of Greek tragedy, so far as I know one of the undeveloped motifs in Hegel scholarship is the relationship between Hegel and Goethe, and the effect upon Hegel of *Faust II*. I was deeply affected by German émigré scholars and their love of *Faust II*, some insisting that it was the Bible of the modern German mind, or at least that it had far more affected that mind than any other literature, and while I may be the only one who thinks this of Heidegger, I do think that theologians must take this work seriously. Are you familiar with Eckerman's record of Goethe's conversations,[4] and the account of how seriously Goethe took Hegel, and was himself affected by Hegel? And his encounter with Hegel took place while he was writing *Faust II*.

I was also intrigued by your taking so seriously the question of the death of art and religion in Hegel, and has anyone investigated the relationship between an Hegelian understanding of the death of art and the death of God, are they two sides of the same coin, and each inseparable from apocalypse itself, or from the advent of absolute actuality? True Agamben is very good on the death of art here,[5] but he does not understand it by way of the death of God, even if his understanding of the death of art is so much fuller than Hegel's. Perhaps one must have encountered postmodernity to be open to an actual death of art, and I do think that Hegel's understanding of the death of art is an understanding of the historical fulfillment of poetry as it occurs in Schiller and Goethe, and perhaps something comparable has occurred in our world, which is how I attempt to understand the fulfillment of epic in Joyce. But your letter induces me to realize how little we have done in this arena, and if there truly are remarkable individual studies here, we have a genuine void philosophically and theologically, but is this due to a deep prohibition, and one of which we are unconscious?

I recognize that this is a most inadequate response to your weighty letter, for I fear that my mind is not fully awake, and there is little so difficult to do as to genuinely think with Hegel, which is what one must do to enter this arena, even if I believe that it has been entered by many, including perhaps the best thinkers of our time. There is a collection of essays that I have not read on Hegel and the philosophy of science, but I did myself contribute to that volume on Hegel and the philosophy of religion,[6] papers from an exciting conference on this subject, one held in South Carolina, and I was the "center" of that conference for I was the only one to bring bottles of whiskey with me, knowing that this area of South Carolina was dry. I wonder if something comparable to this has made me the center of radical theology, could that be because I know more than anyone else, or more concretely than anyone else, that the theological world is "dry." And most dry because it closes itself to Hegel, and therefore ultimately closes itself to modern thinking itself, and must do so simply to be a theological world. So far as I know the only serious theological response to Hegel before mine was Barth's, and that is because Barth could know Hegelian thinking as a far fuller inversion of faith than any liberal theological thinking, but thereby it reveals the deeper consequence of all liberal or modern theology. Of course, I think that there is a deep truth in this, and although I have not thought that there is an integral relationship between Barth and Hegel, there is one between Barth and Nietzsche, even if unrecognized by Barth and the Barthians.

Yours gratefully,
Tom

Letter 28

To Friends (January 24, 2007)

Friends,

In writing my recent paper on American nihilism[1] I spontaneously came upon what I presume or suspect is a new understanding of American religion, and that is that a uniquely American religion is born in response to and as a consequence of a truly new secular world, so that in that context a uniquely American religion is truly distinct insofar as it has no integral relation to historical religious traditions, and is impelled to create its own. This is most manifest in the most uniquely American of all American religions, Mormonism, but to one degree or another is manifest in all American religions, for the birth of American religion is simultaneously the birth of a secular American language and a secular American society, thereby impelling American religions into a new secular mode of life, and one demanding truly new and ahistorical expressions of religion. In this perspective, American Puritanism is truly new, and not only because of its original theocracy, but also because of its deeply secular horizon, a horizon giving birth to that very theocracy, and a theocracy deeply in conflict with that horizon, thus creating a truly new American Calvinism. This is a Calvinism that not even Weber could understand, but it, too, embodies an innerworldly asceticism, a praxis that Weber could know as transforming the world, but American Calvinism is even more secular than its European counterpart, just as its praxis occurs in a far more secular world. Our greatest modern prophet, William Blake, in his first prophetic poem, *America*,[2] could envision the American Revolution as effecting the death of Urizen or the uniquely Christian God, a death ushering in apocalypse itself. Now not only is this the first recorded vision of the death of God, but it is the inauguration of a uniquely modern apocalyptic vision, and one that is truly and ultimately a revolutionary vision.

This is a context that truly illuminates the birth of American religion, a birth that can be understood to be an apocalyptic birth insofar as it is the consequence of the advent of a truly new world, a world more purely secular than any world hitherto, hence demanding a truly secular religion, or a religion directed to the very secularity of the world. Thereby there occurs the birth of a uniquely American praxis, a genuine pragmatism, which America knows as its own unique philosophy, but a philosophy first realizing itself in a truly new American Calvinism, and in the ultimate praxis of that Calvinism. This is a praxis directed to the transformation of the world, and whereas Calvinism had always been directed towards that transformation, only American Calvinism pragmatically embraces that goal, thereby creating a genuinely new process of secularization. What truly distinguishes American secularization from European secularization is that American secularization is simultaneously an ultimate renewal of religion, so that the new secular world of America is inseparable from a truly new religious world, a world with no counterpart elsewhere, for it is a religious world integrally and essentially related to a secular world, and is so even when it is ultimately attacking that world. So it is that it is America that gives birth to Fundamentalism, a fundamentalism far distant from European orthodoxies, for not only is this fundamentalism closed to European historical traditions, but it is inseparable from an ultimate anti-secularism that is only possible for a uniquely American expression of religion, only possible in an American secular world, or an American religious world generated by that new world. So, too, contemporary American politics is religious as is no European politics today, but religious in a truly secular mode, abandoning all genuine religious tradition, and abandoning or suspending all dogmatic language, but doing so in such a way as to be sustained by the dominant public world of religion.

Indeed, the American public world or public realm is itself essentially new, and new as a realm that is secular and religious at once, thereby leading to an abrogation of all established meanings of both the religious and the secular, and if the first separation of church and state occurs in America, this is a separation generating both a truly new government and a truly new church. It is remarkable how little thinking has occurred about the separation of church and state, for while this has occurred in legal thinking, it has not occurred in theological thinking, or there occurred only apologetically or ecclesiastically, never occurring in fundamental theological thinking itself.

Tom

Letter 29

To Friends (January 25, 2007)

Friends,

I wish to share with you a theological awakening that I have received from John Meyendorff's *Christ in Eastern Christian Thought*,[1] one most concretely occurring here in his discussion of Theopaschism[2] in post-Chalcedonian ancient Eastern theology, although Theopaschism originated with Cyril of Alexandria.[3] Just as Cyril insisted upon the title of *Theotokos*[4] because the flesh of Christ is the "flesh of God," this is the flesh in which the Son of God suffers and dies, so that the God made man was born and suffered in the flesh. But the distinction established at Chalcedon between *phusis*[5] and *hypostasis*[6] was too new and revolutionary not to generate enormous controversy, and it occurs particularly in this arena, placing deeply in question the identity of the "flesh" of God. Now an essential aspect of Cyrillian theology is that the Word as the only "subject" of Christ is also the subject of that death in the flesh that "is his own," inevitably leading to Theopaschism, for here it is God alone who could vanquish death by voluntarily assuming it. So it is that the Word's passion in the flesh is the passion of God. Nonetheless the incarnate Word's hypostasis could not have suffered in the flesh since it was a divine hypostasis, and divinity is by nature impassible, as the very Chalcedonian term "hypostasis" inevitably generates profound theological problems, and above all so within soteriology.

Theopaschite expressions are to be found not only in Paul but in pre-Nicene theology as well, and perhaps most forcefully so in Gregory Nazianzen[7] who already makes it the central element in soteriology: "We needed a God *made flesh and put to death*,"[8] just as he also uses the expressions "blood of God" and "crucified God." Leontius of Jerusalem[9] later establishes an absolute distinction between hypostasis and nature, wherein the Word remains impassible in his divine nature, but suffers in his human nature. Since from the moment of the Incarnation the human nature had become as fully his own as the divine nature, one must say that the Word suffered hypostatically, in his own flesh, because his hypostasis is ontologically distinct from the divine nature, and "the Word suffers according to the hypostasis, for within his hypostasis he assumed a passible essence besides his own impassible essence, and what can be asserted of the passible essence can be asserted of the hypostasis."[10] In this way the new formula "one of the Holy Trinity suffered in the flesh" clarifies the Chalcedonian formula, and for Meyendorff all of this theology is grounded in a deeply Eastern Christian deification of humanity, one originally occurring in the deification of the hypostasis of Christ.

All too significantly this theological movement reaches a decisive fulfillment in the reign of the most powerful of all post-Constantinian emperors, Justinian, and even does so in the *Confessions of Faith* composed by Justinian in 551 and addressed to the whole fullness of the Catholic Church. The Emperor strongly asserts the orthodoxy of the theopaschite formula, which had become for him a real criterion of orthodoxy, and does so in the context of Chalcedonian orthodoxy, then proceeding to give 14 anathematisms that the Fifth Ecumenical Council of 553 renews, anathematisms wherein the unity of Christ is designated by the notion of hypostasis and not that of nature, and the nature can exist only *within* the hypostasis, for otherwise it remains an indeterminate abstraction. The Fifth Council by fully rehabilitating the Cyrillian concept of the unity of the *subject* of Christ, focused attention upon the great importance of the hypostatic unity of the incarnate Word, and in conformity with Chalcedon Christ is fully consubstantial to us by his humanity, but that his hypostasis was the divine and preexisting Word.

The decisions of the Fifth Council for Meyendorff do not constitute a final conclusion but only a stage in the development of Christology, and their dogmas must be considered in the light of later stages, and especially of St. Maximus's doctrine of the two wills and his conception of deification. And critics of the Christology of the Fifth Council seem to Meyendorff to base their judgment on the Thomist concept of "pure nature," which cannot be reconciled either with the patristic conception of sin of with that of deification. Deified human nature, human nature having come into communion with the divine nature, is not "modified as to its natural characteristics," but restored to the divine glory to which it was destined since the creation. Human nature, at the contact of God, does not disappear, *on the contrary it becomes fully human*, for God cannot destroy what He has made. The fact that the human nature of Jesus was hypostatized into the Logos does not suppress that human nature but, on the contrary, guarantees its own human perfection and therefore its consubstantiality with the whole of the human race. And this is the human race that is redeemed by the Passion of Jesus Christ, a passion that is the Passion of God, and a passion of God that *is* the suffering and death of God.

Is it possible that all of this is a fundamental reason why I have never encountered any opposition from an Orthodox theologian but rather blessing and peace?

Tom

Letter 30

To Friends (January 26, 2007)

Friends,

I have now returned to Harnack's *History of Dogma,* which I only recently reread to renew my understanding of what Harnack refers to in the fourth volume as the so-called Theopaschitian controversy. The formulae "God has suffered," "God was crucified," were time-honored forms of speech in the Church and had never been quite forgotten, and after the formulation of the Chalcedonian Creed, it seemed to be impossible to disapprove them, for if Mary were to be called Theotokos this meant that they were approved. Nevertheless opposition occurred, and one of a quite complex nature, and a controversial formula arose about 518, "one of the Trinity was crucified—suffered in the flesh." The legates of Pope Hormisdos[1] opposed it, and the Pope himself became concerned, but only with great ambivalence. Actually the issue was decided by Emperor Justinian, who then impelled the popes to acknowledge it, and then got the formula sanctioned by the Fifth Ecumenical Council that "our Lord who was crucified in the flesh, Jesus Christ, was one of the Trinity."

Now Harnack contends that all of this led to a "complete change" in the theology of that period, and the crucial figure is Leontius of Byzantium (485–543), the first scholastic, or first Aristotelian theologian, one loyal to the Chalcedonian formula, but who interpreted it through Aristotelian categories. Leontius transferred the Christological controversy into the philosophical realm, and this for Harnack is a momentous revolution! It is undeniable for Harnack that Chalcedonian orthodoxy was first firmly established in the East in the age of Justinian, and for Harnack the only possible explanation for this is the entrance of Aristotelian scholasticism into the Church, and that means for him that the Chalcedonian dogma is lost in philosophical theology. Leontius was a Scythian monk who was a great opponent of the Monophysites and who championed the Theopaschitian formula, but his Aristotelianism called for a restatement of the doctrine of the Trinity, and the "unus ex trinitate" is opposed to the "thrice holy" who was crucified for us.

Accordingly, for Harnack, it is only when dogmatic theology becomes philosophical theology that the Crucified God becomes fully understood and sanctioned by theology, and, while unmentioned by Harnack, there is a remarkable parallel to this in Hegel's philosophical revolution!

Tom

Letter 31

To Lissa McCullough (March 13, 2007)

Dear Lissa,

Recently I have been thinking about the writing of the birth of vision chapter in *History as Apocalypse*,[1] identifying it as a crucial time of breakthrough for me, and wondering just how this was related to my initiation by Mircea.[2] This chapter was initially inspired by my encounter with Wendy Doniger[3] that occurred on the Friday evening before the Saturday morning of the Eliade initiation, and this was the time when he informed me that he had chosen Wendy to be his successor, which came as a genuine surprise to Wendy when I communicated this to her. You will remember that it was after I left Wendy's that I encountered the "ghosts" with whom I danced in the streets and after whom Mircea inquired the next morning.[4] Now I deeply remember that it was while writing this vision chapter that I was overwhelmed by an inspiring Spirit that I had never encountered before and that I was given a vision that I had never imagined before. True, fundamental breakthroughs had occurred before, but these occurred by way of an immersion in primal texts as opposed to the immediacy of this experience. And it was an "experience," and an all too actual once, and an experience that was an immersion in the very body or bodies of the Goddess or the goddesses. Nothing prepared me for this, and certainly no theological thinking did, indeed, the very opposite is true, and yet perhaps this was necessary for my voyage into the Christian epic.

Although the literature on Joyce is vast, and I only know a fraction of it, I nonetheless have the sense that this is the only book that fully draws forth the role of the Goddess in Joyce, just as it is apparently alone in identifying Dante's Beatrice as the incarnate Body of Christ, and almost certainly alone along with my book on Blake in apprehending Blake's Jerusalem as a *coincidentia oppositorum* of Christ and Satan. Now it is also true is it not that this is the only book in which I genuinely explore the Goddess and actually explore the Body of the Goddess?

Even the section on Leahy here calls attention to the absolute role of the Mother of God in his *Novitas Mundi*,[5] and this was so strange to me that I had to consult him in my wording of this role. Now simply speaking of the book in this manner is a way of realizing the necessity of its first chapter, a necessity that has commonly been alien to me, and surely this chapter stands wholly alone in my writing, or is that not true?

But if it is true what is the significance of this?

Sometimes I think that Catholic theologians can only rarely think either radically or genuinely about the Mother of God, and it is certainly a profoundly forbidden topic in Protestant theology, just as it appears to be absent from the whole world of genuinely modern theology. Yet the Mother of God is truly primal in the Christian epic tradition, being absent only in Milton, but that is an essential absence for Milton's revolutionary discovery of the totality of Satan, a totality here dialectically inseparable from the totality of Christ. So, too, it could have been essential that Milton's rebirth in Blake, and dialectical rebirth, necessarily issued in a rebirth of the Goddess and goddesses, and one even going beyond their ultimate role in Dante. And is it true that only the death of God makes possible such a rebirth of the Goddess, a death of God forever ending that Law of the Father that represses the Mother, a Law that perhaps Lacan has known most deeply in our world?

In this perspective, my subsequent work has moved very far from this ground, and it is difficult to see how it could have been grounded in this work, or can you see this? Of course, all of my later work is grounded in the Christian epic, but is it open to the actual Body of the Goddess, or to body at all? I know that you and others are truly critical at just this point, but is a closure to the body here at bottom a closure to the Body of the Goddess? And could this also be a closure to a fundamental ground of my own theological breakthrough, one perhaps denying its most fundamental ground, thereby making impossible a resolution of my own work? These are questions that are beginning to obsess me, and while this may soon pass, are these genuine questions for me, and finally inescapable questions?

Yes, night must fall, and my night must soon fall, but I shall not go gently into that night, and will continue to struggle so long as I am open to life.

Yours gratefully,
Tom

Letter 32

To Donald Weinstein* (March 23, 2007)

Dear Don,

If only in response to now being in communication with you I have begun again to muse about America and religion, and to attempt to think about the uniqueness of American religion, since I have little confidence in any of the historians of American religion whom I have read apart from Perry Miller.[1] Some years ago I was much affected by the idea that a new kind of establishment occurs with the birth of America, one in which not only a new government but a new society is established, and that this occurs not simply in the context of the Enlightenment but occurs as the birth of a truly new secular society. Although I initially resisted Hannah Arendt's[2] understanding of the American and French revolutions, I nonetheless am drawn to the idea that few of us yet understand the American Revolution as a genuine revolution, and that a major reason for this is that we have a conservative understanding of American religion, and are commonly closed to its genuine uniqueness. As the years have passed by I have more and more thought of myself as an American theologian and as a uniquely American one, and above all so as an atheistic theologian, since there are virtually none in Europe. I also have to acknowledge my Calvinistic heritage, if only through Stonewall Jackson,[3] and it is true that I am a theologian of predestination as none has been since Barth, and while writing my memoir I came to see how primal predestination has been for my theological thinking.

But the one who most deeply influenced my understanding of predestination was Max Weber, and in Weber we can understand how an ultimate predestination and an ultimate secularization go hand in hand, for this is a predestination that is inseparable from the historical advent of an innerworldly asceticism, and Weber understood this as occurring at the first time in history in which there was an unprecedented inner isolation of the individual. Yet what Weber missed was that this first corporately or nationally occurs in America, and in the birth of an American society and an American religion in which for the first time both society and religion have broken away from a living or continuous historical tradition, for we Americans truly are alienated from history, as most concretely manifest in the most distinctively American expressions of religion. For I am coming to believe that a fully corporate secularism was only born with the birth of America, just as a fully secular society was then born, and that a genuinely American religion is hence inevitably a genuinely secular religion, or is so in its deeper even if invisible ground. Or, stated more critically, American religion is born in the context of the ending of Christendom, an ending that was not even understood until Kierkegaard, but that ending entails the necessity of a new subjectivity of faith itself, which Kierkegaard so deeply understands, and such a faith is born with the advent of American religion. Yet as Kierkegaard understands so profoundly, such a subjectivity of faith is inseparable from its ground in an absolutely Godless world or "objectivity," hence the necessity of Kierkegaard's Hegelianism, a necessity truly paralleled in Marx, and it was Kierkegaard and Marx who most purely understood Hegel as an ultimately atheistic thinker.

Now I think that it is also true that American religion is the craziest or most pathological religion in the world, but this is essentially related to its deeply secular ground, and a ground itself inseparable from an unprecedented inner isolation of the individual, an isolation inevitably generating deep pathologies. Incidentally, those BBC tapes[4] I sent you today are directed to demonstrating or showing how psychoanalysis was employed by government and major corporations to control or pacify the American public, an initiative originated by a fear in high circles of a potentiality for chaos in America. If American society more than any other is obsessed with psychotherapy, this is not unrelated to a uniquely American condition, and a truly pathological one, but one only openly expressed in American religion, which has surely been a driving force in both the Vietnam and the Iraq wars.

In other words, I think that a new Calvinism was born in America, and a truly individual or subjective Calvinism, one inseparable from a profoundly secular ground. This is a Calvinism that at bottom is a truly churchless Calvinism, for if it is not until the advent of America that a sectarian religion becomes all in all, which is an inevitable consequence of the first full separation or isolation of church and state, such sectarianism is inseparable from its truly and fully secular ground. Many years ago I was much affected by Mencken's book on the American language,[5] and by the force with which he establishes the uniqueness of the American language, the first truly secular language in history. Now I know that you resist all such historical claims, but I am an Hegelian and Nietzschean at once, and just as I know Hegel and Nietzsche as our greatest historical thinkers, it is Hegel who in the conclusion of his lectures on the philosophy of history can call forth America as the land of the future, and that was and is a radically secular future, and one already at hand in the advent of America, which itself at this crucial point is a truly revolutionary advent.

Somehow you have impelled these musings, or, at least impelled them today.

Tom

Letter 33

To Friends (March 27, 2007)

Friends,

I have just begun to reread Marcel Gauchet's *The Disenchantment of the World: A Political History of Religion*,[1] and I wish to share with you statements from its introduction. He begins by stating his famous thesis that the religious has reached the end of its life in the modern world, but then the modern Western world's radical originality lies wholly in its reincorporation of the religious into the very heart of human relationships and activities, this is the reincorporation of the sacred element, which previously shaped this world from the outside. Religion's demise is to be ascertained by the extent of this restructuring, and though it originated within religion, it escaped from and reversed its original religious orientation, and this reversal is this book's main focus.

Christianity now proves to be a religion for departing from religion. But the age of religion dominated virtually all of history, and we have only broken away from religion by finding substitutes for it at every level, and here the primary concern is to recover the significance and coherence of this basic human option. A chasm separates us from our origins, and the most obvious difference between us and our origins is their lack of a State, while the main similarity lies in the role of religion. Here, the emergence of the State is the main event in history, and this corresponds to a massive revision of the human situation, to an absolute transformation. And while all the elements in the early arrangement will be found in the later one, now their embodiments in the world have nothing in common. After dozens of millennia with religion dominating politics, followed by five millennia with politics dominating religion, we are at the point where religion has been systematically exhausted and its legacy has gradually disappeared.

Indeed, the so-called major or universal religions, far from being the essential embodiment of religion, are in fact just so many stages of its abatement and disintegration. And the greatest and most universal of them, our own, is precisely the one that allows a departure from religion. When dealing with religion, what appears to be an advance is actually a retreat. Here, Gauchet's deeply French roots are manifest, as Rousseau, Durkheim, and even Comte return, but he also brings an enlightening perspective to that radical French thinking that occurred after World War II, one in which a radical deconstruction is finally an opening to a purely archaic or purely primordial ground. Despite his adoption of Weber's language in the title of the book, he is vastly distant from Weber, and distant, too, from the German philosophic tradition, and perhaps more distant than any other contemporary French thinker.

Gauchet is a professor of social science[2] but he is wholly unlike what we know as anthropologists, for whereas he has a primary interest in primeval societies, he knows these as societies prior to the State, hence they are invisible today, and not subject to empirical investigation, and he can point to Africa as a region vastly distant from primeval society. Now in societies prior to the State, religion does indeed play the central role, and this system entails the radical anteriority of every ordering principle, and is thus a system of dispossession, inheritance, and immutability. Nothing we behold or do in everyday life derives from us, but rather from a different species. Our only function is to preserve their inviolable legacy and to repeat their sacred teaching, and this is the genuine logic of religion, and absolute separation from the original model is the vector of its absolute dictatorship. All of this sounds like a parody of Eliade, but Eliade is unmentioned in the book's bibliography.

For Gauchet, the emergence of the State is the first religious revolution in history, and it contains a strictly spiritual revolution, which redistributed the terms of the original arrangement by incarnating among humans what had been separated from them. What was originally excluded became embodied within society, so that from now on there will occur the systematic subversion of religious life. Yet there are two major paths: one compromised between maintaining the original religious structure and integrating new material—the path of oriental religion with the notion of being as a void: the other, directly opposed to it, was the radical path of subjectivizing the divine and structurally separating the physical from the spiritual. Though destined to turn the world upside down, this later path, Jewish monotheism, was initially extremely marginal, but when Christianity takes it over the seeds of its terrestrial autonomy come to fruition. This is the point, where thanks to religion, a society with no further need for religion arises.

Now to judge by his bibliography Gauchet is wholly disinterested in the disciplines of anthropology and the history of religions, nor is he open to biblical scholarship, and he declares at the conclusion of his introduction that Montesquieu or Rousseau, Tocqueville or Marx still tell us more about society than all of sociology combined. Hence Gauchet is a thinker rather than a scholar, or as the book jacket asserts he has launched a speculative history, but a speculative history that is apparently having an enormous impact.

Tom

Letter 34

To Friends (May 7, 2007)

Friends,

I continue to be enchanted by Marcel Gauchet's *The Disenchantment of the World: A Political History of Religion*, and not only enchanted but mystified for this is not only a sociological-historical book but also a truly theological book, and I mean theology in a fully dogmatic sense. Indeed, I think that one must accept this as a genuinely Christian book, and Christian even in the confessional sense, for Gauchet (like Hegel!) knows the fullness or the fullest actuality of history as a Christian actuality, and the actuality of our world itself. For if Christianity proves to be a religion for departing from religion, that departure is a consequence of the Incarnation itself, and even of the actual way or praxis of Jesus. How fascinating that a truly major and seminal historical-sociological work could so deeply draw forth both Jesus and Christology, and this in its second part on "The Apogee and Death of God," and, yes, Gauchet, is a death of God theologian, and even an apocalyptic theologian. Perhaps the French didn't translate my work because they didn't need it!

Thus Gauchet can claim at the beginning of the second part that the wholesale reconstruction of human space under the influence of God's paradoxical "absolutization/withdrawal" is the hidden source behind the expanding fragmented components of our democratic, individualizing state-based, historical, technological, capitalist world, which seem contradictory but are essentially unified. For we originated in religion, and continue to explain ourselves through it, and always will! Now the entire historical process leading up to Christianity, including the Incarnation, should thus be understood in terms of the spiritual reversal of the system of universal power, as Augustine's *City of God* is here reborn, although here there is a far more total celebration of the *presence of the transcendent*. And that presence fully occurs in that Jesus who is the inverted Messiah, one whose way realizes the ultimate union of heaven and earth by announcing the imminent arrival of universal upheaval, an upheaval that will reverse all possible mediation between heaven and earth, but nonetheless realize that truly universal Godhead that is accessible to all. For Jesus embodies something unutterably new, an absolute novitas that will wholly transform everything whatsoever, thus ending religion itself, but that ending is an absolute beginning, and an absolute beginning of a total presence.

Who could imagine a sociological-historical work making such claims? And why has Gauchet had such an enormous impact? Is theology now far more fully alive that we theologians can realize?

Tom

Letter 35

To John D. Caputo*
(May 17, 2007)

Dear Jack,

Recently if only indirectly I have been responding to you, and doing so by way of thinking about radical Catholicism, a topic that I have long taken seriously but I do not remember our having fully discussed it. My favorite examples of such a revolutionary Catholicism are Joyce and Leahy, but Leahy is truly paradoxical in being simultaneously more orthodox than the Pope and an ultimately radical thinker in thinking the crucifixion of God as apocalypse itself, which, of course, is a major motif of *Finnegans Wake* if not of *Ulysses* as well.

Many years ago I tried but failed to induce Tom Sheehan[1] to bring together an anthology of contemporary radical Catholic thinkers but had little effect unless that is occurring today, and I deem it significant that both of you are radical Catholics and Heidegger scholars at once, for I want to understand Heidegger as a radical Catholic. Of course, I am not a Heidegger scholar, but I may be more open to him theologically than you and Tom are, and I see Heidegger as being in continuity not only with Schelling and Nietzsche but with Hegel, too, and I don't think that his theological thinking can be understood apart from this. Now it is fascinating how much is being published on Heidegger and theology today, but I am most affected by the discovery that Heidegger is a genuine poet, as revealed in the Arendt correspondence. So far as I know it is only in this correspondence that he can speak with genuine intimacy, and while there is little that is overtly theological here, nonetheless a genuine theological horizon is clearly manifest. That is what is most fascinating to this theologian, for nothing is more difficult for theologians than to establish a genuine theological horizon, which here occurs so spontaneously.

I think that most are unaware of how fully theological both Arendt and Kristeva[2] are (Kristeva once asked me to think of her as a French death-of-God theologian), and this is because we commonly have such a narrow understanding of theology, one sanctioned by the majority of theologians if only to preserve this turf for themselves. Clearly contemporary writing on Heidegger and theology is challenging such an understanding of theology, and even if few theologians respond to this, a new way is nonetheless being called forth, and one that excites me. I now much regret that we withdrew from our theological dialogue, and even if this was occasioned for me by my terrible back pain,[3] that problem has now been largely solved by regular exercise in a gym, thus releasing me as I have not been released for some time. While I do have writing projects under way, for example a book on the apocalyptic Trinity, various blocks have occurred, and above all a block on a projected book on theological ethics as I ever more fully realize that I simply cannot write explicitly on ethics, even if I am not ashamed of the chapter on ethics in my memoir. I am also persuaded that a powerful ethics can be hidden or invisible or disguised as in Heidegger, or seemingly wholly perverse as in Nietzsche, or apparently simply pagan as in Aristotle.

Now the very question of atheism is truly explosive in Hegel, Nietzsche, and Heidegger, and this is the context in which I respond to your recent claim that Levinas himself is an atheist,[4] just as I sense that this is a perfect example of your truly paradoxical and "offensive" thinking, employing the word *offense* in a Pauline and Kierkegaardian sense. Among philosophers only Nietzsche has been a profoundly offensive thinker, and even if this is just what Heidegger misses in his understanding of Nietzsche, Heidegger thereby misses Nietzsche as a profoundly theological thinker, although this may well be only Heidegger once again disguising himself. Think of how long he got away with disguising his own Kierkegaardian ground, a disguise essential to his theological mask, and if he refused to publish *Beitrage* while still alive, is this, too, not a theological mask? By the way, I cannot remember (age is affecting my memory) if I asked you about Bill Richardson's essay on *Beiträge,* which excited me; this was supposed to become a book but I have heard nothing about it.[5]

Is Heidegger not already a genuine Catholic thinker in his thinking about death in *Being and Time*? Yes, I know that he had long since abandoned or withdrawn from the Church, but just what kind of withdrawal is that, could it be one essential for his own genuine but radical Catholicism? We must remember that during this period radical or even genuine thinking was forbidden by ecclesiastical authority, and forbidden for a very good reason, for it seemingly could only be a truly heretical thinking. The topic of death is a good example of this, for it cannot be truly thought within a scholastic context, and not even by Augustine himself. Or, rather, Augustine can only genuinely think about death in thinking damnation itself, and if he is our deepest or purest thinker of damnation, at no other point has he been so overwhelmingly influential, even if both our philosophical and our theological worlds have become closed to damnation. But does not the ultimate power of Heidegger's thinking of inauthentic existence derive from its hidden horizon within an eternal damnation? And just as there is a truly Pauline ground in Heidegger's ultimate affirmation of death, that is a ground inseparable from an absolute dichotomy between eternal life and eternal death, but it was the very realization of that dichotomy that made possible Paul's discovery of self-consciousness itself, a self-consciousness realizing its philosophical consummation in *Being and Time*.

I think that for many it is very difficult to understand Heidegger as a Catholic thinker because he is a genuine Pauline thinker, but it was Augustine who philosophically discovered Paul, and just as Augustine is our first truly Pauline thinker, he is also as purely a Catholic thinker as anyone, and if only in Augustine a purely Pauline thinking is a purely Catholic thinking, one that becomes impossible in modernity, as witness Jansenism. This is the context in which we must approach Heidegger as a Catholic thinker, for a Pauline Catholicism is now either forbidden or impossible, and just as this condition is inseparable from a uniquely modern nihilism, that is a nihilism that Heidegger can know as being embodied in the very history of Being, thereby Being becomes wholly negated, forgotten, or eclipsed, and it is just here that Heidegger is most distant from Catholicism, or from a uniquely modern Catholicism, and one already born in Aquinas himself. Yet is it not possible to understand an ultimate distance from modern Catholicism as making possible a genuine renewal of Catholicism, but a Catholicism that can only be a radical Catholicism, and most manifestly radical in its vast distance from all possible scholasticism? Is this not true of Joyce, and perhaps of every truly major modern Catholic voice, a voice that is inevitably heterodox from every given or established Catholic perspective, and not only heterodox but truly atheistic, or truly atheistic in the perspective of the orthodox Catholic God. This is how I can understand Heidegger to be an atheist, but I am nonetheless persuaded that Heidegger thinks God more purely than any other twentieth century thinker, and that this is the only thinking of God that has been real in a truly nihilistic world and horizon.

As you know I am intrigued that Benedict XVI has such an astute and comprehensive understanding of nihilism, even if he can only publicly speak of it as relativism, and if he can now affirm that it is Protestant evangelism that is the deepest threat to Catholicism, is this not an evangelism that is yet another expression of a uniquely modern nihilism, one ending not only thinking but all possible expressions of either the imagination or a deeper life? Nonetheless Benedict XVI has been at the forefront of a repressive movement assaulting all truly modern or contemporary Catholic thinking, and while I can understand the ground of this if only because we truly exist in an atheistic and nihilistic world, a genuine Catholicism simply cannot be wholly bound to the past, for that forecloses all genuine movement into the future. Hence it is only radical Catholicism that can now actualize such a movement, and here you have been given an ultimate vocation, and one that I deeply respect.

I want you to know that my present thinking is not truly distant from yours, and that I continue to honor your life and work.

Tom

Letter 36

To Brian Schroeder (June 7, 2007)

Dear Brian,

I have just concluded your excellent essay on apocalypse and eschatology,[1] which is perhaps the most decisive way of relating Nietzsche and Levinas. However, I think that a misstep initially occurs that is ultimately misleading to your purpose, and that is your definitions of eschatology and apocalypse by way of an etymological analysis of the Greek sources of these words, whereas as you later note these are alien to Greece and originate in Israel. Moreover, I think that you miss something that is crucial to prophetic eschatology, as so profoundly understood by Weber, and that is its absolute challenge to the Given, the most ultimate challenge that had thus far occurred, and one inseparable from a total assault upon an old Israel or even an old covenant, as concretely occurring in a total negation of monarchy. This is why I have always believed that messianic prophecy is alien to prophetic eschatology, and is a consequence of priestly Israel, for in messianic prophecy as such the Messiah is a sacred or ultimate Monarch, which is precisely why Jesus could not have identified himself as the Messiah. Is the Jew as Jew inevitably drawn to messianic language, as witness both Levinas and Derrida, but perhaps Kafka and Benjamin are here deeper Jews, and this is your own approach in discussing Judaism, where you apparently are audacious enough to challenge the Judaism of Levinas.

Again I think that you allow Levinas to mislead you about Hegel, for Hegel is our first apocalyptic philosopher, and is so by being the first philosopher of the death of God. Nor is the death of God here simply a conceptual realization, for it is first realized in the depths of the Unhappy Consciousness, and even gives birth to an ultimate *Angst*, and even if that *Angst* is destined to an ultimate reconciliation, such reconciliation brings an end to everything that we have actually known as both history and consciousness. Here, Nietzsche is deeply Hegelian, even if he knew very little Hegel, but here, too, both Hegel and Nietzsche are ultimately distant from Levinas, as all too well known to Levinas himself. You are also audacious in making the death of God so important to Levinas, who is surely extremely reticent in discussing it, for he knows all too well that it is not simply a certain conception of God that has perished, but far rather everything that has been manifest as God in the Western tradition itself, which is yet another reason to distance Judaism from that tradition.

I think that you are deeply right in so ultimately distancing Levinas from apocalypticism, and this does make possible a distinctively Jewish or Hebraic eschatology, one giving birth to the *illeity*[2] of Levinas, which is perhaps his most original and important thinking, but one that can only be understood theologically, which is why it is so weakly grasped by Levinas scholars so far as I know. Is it possible that Levinas is our most important eschatological thinker, and is so by being a purely anti-apocalyptic thinker, thus making possible an anti-apocalyptic eschatological thinking, which is perhaps the first such thinking to occur, unless this has always finally been true of an authentically Jewish thinking? And does this mean that Jewish thinking can only be Jewish thinking by being an anti-Christian thinking? Remember that Jacob Neusner understands the birth of Judaism as being a negative reaction to the birth of Christianity, for despite all common understanding, Christianity existed before Judaism, even if only just before Judaism.

I do think that your conclusion is all too brief and is in much need of expansion, if only to complete the essay itself, and here the death of God must come to the fore once again, perhaps eliciting an ultimate distinction between an eschatological and an apocalyptic death of God.

Yours gratefully,
Tom

Letter 37

To John T. Wilcox* (June 13, 2007)

Dear Jack,

Let me first respond to your remarks about Nietzsche, for I fear that a deep division lies between us here, and one that cannot be resolved by argument. For I cannot believe that Nietzsche's collapse was simply due to an organic condition or to syphilis, not only because his own late thinking was so profoundly centered upon that absolute ending that is an absolute transfiguration, but as he himself knew so deeply it was courage above all that this venture entailed, and a courage braving those ultimate depths that finally destroyed him. Here, I like to employ the image of the trapeze flyer, whose ultimate trick must be done without a net, as powerfully portrayed in one of my favorite movies, *Trapeze*.[1] Nietzsche ultimately thought without a net, and he paid for it profoundly, as such he is a unique thinker, but there are poets such as Hölderlin[2] who parallel him here, and perhaps Blake as well, and it is not insignificant that there are such profound parallels between Blake and Nietzsche.

The Book of Job presents a very different kind of problem, even if I think that Nietzsche all too indirectly is the most profound interpreter of that book, although I would first respond to you by saying that I accept an historical-critical approach here, so that these different perspectives or positions of the book derive from different sources or grounds, which themselves are incompatible with each other. First, however, we must acknowledge that this is the most heterodox or dangerous book in the Bible, and the miracle is that it was ever canonized, even if there are many other heterodox writings in the Bible (a fundamental reason why the fundamentalists can never be honest literalists). And I interpret the theophany or the Voice from the Whirlwind not as the genuine conclusion of the book, but rather as its ultimate antagonist whom Blake in his illustrations to the Book of Job named as Satan, that is to say that this is that God who is the absolutely sovereign and alien God whom Melville named as Moby Dick, and Hegel knew as the "Bad Infinite" or purely abstract Spirit. So that for me the center and body of the Book of Job are the dialogues between Job and his friends, truly dramatic dialogues, the most dramatic writing in the Bible, and dialogues moving ever more fully in the direction of naming the absolutely transcendent God who is the absolutely alien God, and that here is possible only by way of reversing the orthodoxy of Job's friends. The ultimate struggle between Job and his friends is the most dramatic conflict in the Bible, and just as many know this as the greatest literary language in the Bible, it is one in which Job's language becomes ever more deeply human, and his friends ever more deeply alien, but it is Job's most human or most intimate or most "existential" language that names the Alien God, and yet this is a naming that has been profoundly resisted by virtually all of our traditions.

I find the nineteenth chapter to be the most challenging one in the book, the one point at which Job can name his redeemer, and also declare that he will see God by his side, but surely this can occur only by way of a reversal or transfiguration of the Alien God, an alien God that must pass into grace and grace alone, which I affirm to be the ultimate motif of this truly heterodox book. Of course, this is a Christian interpretation, and one that does not shame me, but rather makes possible my most radical theological thinking, which is centered upon crucifixion-resurrection as the self-saving of God. This is what I understand as the atonement, which is an atonement of Godhead itself, and whose clearest biblical source is the Book of Job. Unfortunately, I have not been able to center upon this in my writing, coming closest to it in *Godhead and the Nothing*, but perhaps I should take up this challenge now.

It fascinates me that the Book of Job is so deeply loved, perhaps the only biblical book that is truly loved by the secular mind, and yet it is profoundly alien to Christian theology, except for that radical Barth who so profoundly knew all natural or philosophical theology as a truly demonic theology. It is also remarkable that the Book of Job is so absent from the New Testament, perhaps being present only in Paul's profound understanding of the Judgment or Wrath of God, but for Paul that judgment is inseparable from God's grace. Hence the Book of Job can be a way for us into an absolutely mysterious grace, or an absolutely elusive redemption, one that I am persuaded Nietzsche deeply understood, even if that understanding finally broke Nietzsche himself. Writing this has impelled me to glance again at *Godhead and the Nothing*, and I conclude this letter by quoting a paragraph from it if only as a means of inducing myself to return to this motif, and the subject here is the biblical transformation of primordial sacrifice.

"Deep traces of this ultimate transformation are recorded in the Bible, and perhaps most clearly in the Book of Job, surely the most radical and heretical book in the Bible. Even if the original text was deeply transformed by priestly editors, it remains inexplicable how this book could have become canonical. The Creator or El Shaddai of the great bulk of the Book of Job is the God of absolute power and absolute power alone, or that God who is Creator and only Creator, wholly divorced from everything that Israel knew as hope and redemption, and therein divorced from the covenantal and legal traditions of Israel. Only this divorce makes possible the final epiphany of the Creator to Job. All too significantly now the name of the Creator is Yahweh, but a Yahweh deeply isolated from everything that Israel had previously known as Yahweh, as this Yahweh appears out of the deepest and most ultimate void. Now every possible myth and ritual is annulled. While this is a genuine resolution of the deep tragedy of Job, it is wholly divorced from the dramatic dialogues of the book, just as it is divorced from every possible human response, except the response of absolute submission, a submission that is an ultimate silence, and the only such silence recorded in scripture. But this is a silence inseparable from an epiphany of absolute power, as the liturgical movement of sacrifice is now transformed into its very opposite, and the act of God can now be known and realized as an act of absolute power and absolute power alone."[3]

Yours gratefully,
Tom

Letter 38

To Friends (July 10, 2007)

Friends,

Responses to my recent letter on Hegel and theology have induced me to reconsider a fundamental problem, and that is whether or not there has been a response to Hegel's actual theological thinking. While there has been a great deal published on philosophy and theology in recent years, I fear that little or nothing of this is from the perspective of theology itself, and certainly Hegel's theological thinking has not been explored from a contemporary perspective. Indeed, I am coming to look upon Hegel's theological thinking as virgin territory unknown to all, even while looking upon Hegel as the greatest and most creative theological thinker who ever lived. There are also important parallels with Heidegger here, both began as theological thinkers, and as radical theological thinkers, just as each was in quest of an original and lost ground of Christianity. New Testament scholarship would teach us that the name of this ground is "Kingdom of God," one that was the very center of the praxis and proclamation of Jesus, but was almost immediately wholly transformed by the Hellenistic Church. Such scholarship was alien both to Hegel's and to Heidegger's theological training, but each in their own all too distinctive way realized a parallel expression of it, and for Hegel this is embodied in his deepest and most profound thinking, his revolutionary kenotic thinking, in which for the first time the very depths of the New Testament pass into pure thinking itself.

Now I recognize that unlike Hegel and Heidegger the contemporary philosophical world is wholly alienated from the New Testament, just as New Testament scholarship is alienated from the philosophical world or worlds, and perhaps at no other point is there a greater distance between our world and the world of our fathers. The very word *kenosis* is illuminating here, a truly primal word for Hegel, one unquestionably derived from Paul, but at this absolutely fundamental point virtually the whole world of theology has been alienated from Paul, and this is a profound alienation that deeply affected both Heidegger and Hegel. J. N.

Findlay remarks in his book on Hegel[1] that Hegel is alone among philosophers in his purely philosophical realization of the depths of Christian dogma, realizing Creation, Covenant, Incarnation, Crucifixion, Resurrection, and Apocalypse in a new and revolutionary philosophical thinking, when for the first time theology and philosophy wholly coincide or are fully and totally united. Both our philosophical and our theological worlds ignore this extremely important truth, and this alone could account for their mutual alienation from Hegel, an alienation now extending to those profoundly Hegelian thinkers, Kierkegaard and Marx, for even if they are wholly reverse or inverted Hegelian thinkers, that inversion or reversal is inseparable from their revolutionary impact.

It is well established that self-negation or self-emptying is the very essence of a uniquely Hegelian thinking, one that is not only the fundamental source of that thinking, but the source of the forward movement of history itself, not only making possible a uniquely Hegelian historical thinking, but also a fundamental historical thinking truly alien to our world. Heidegger is perhaps most Hegelian in his deep historical thinking, even if that thinking refuses the forward movement of history, a refusal here arising from a post-Hegelian realization of the death of God, and one inaugurating a total and all comprehending abyss.

While such an abyss is truly alien to Hegelian thinking, and here there is a genuine ground for the anti-Hegelian thinking of our world, it is the ultimately radical Hegelian thinking of Kierkegaard and Marx that first and most decisively realizes that abyss, and if only here Marx and Kierkegaard are inescapable for us all. Yet here there is also an inescapable theological ground for us, an abyss that is an ultimate and absolute abyss, and if only thereby a theological abyss, an abyss that has only been fully named in an Hegelian language, even if in a purely reverse or inverted Hegelian language.

This is a language that is a truly possible theological language for us, and this at a time when all manifest theological language is a deeply conservative language, including the language of our seemingly most radical philosophers, yet it is nonetheless true that the call of a radical theological language is now universally ignored. Does this go hand in hand with a deep turn away from Hegelian thinking, and has there ever been a genuinely or deeply Hegelian thinking that is not a radical theological thinking, or not an embodiment of a uniquely modern realization of the death of God? This is just the point at which we are most alienated from an idealistic Hegelian thinking, a late modern idealism alien to Hegel himself, or alien to the Hegel who is a profoundly kenotic thinker. Yet this is the Hegel who is most openly a theological thinker, and most so in his very understanding of God, that uniquely Christian God who is now first known as an absolutely kenotic or self-emptying God, as for the first time the Crucified God is known as Godhead itself. How fascinating that this never openly occurs in almost two thousand years of Christian theological thinking, true, Luther profoundly apprehends it, and if this made possible the Protestant revolution, it could never be formulated in a Protestant dogmatics, and while Barth perhaps attempted it, he only realized it in his doctrine of election or predestination wherein alone he can most fully and most decisively know the Crucified God.

Theologians have long maintained that the Crucified God is truly alien to philosophy, but that is manifestly untrue of Hegel, and profoundly untrue of the deepest and purest Hegelian thinking, or that thinking occurring in the *Phenomenology of Spirit* and the *Science of Logic*. And if this thinking is inseparable from a profound and irreversible enactment of the death of God, that death can here by known as a realization of the Crucified God, and a profoundly philosophical realization of that death, one revolutionizing philosophical thinking itself. Yet, so, too, does it revolutionize theological thinking, and all too ironically so, as the Crucified God is first fully and decisively realized theologically, and that Crucified God who truly and fully is the uniquely Christian God. Now if Hegel first realizes both philosophically and theologically the uniquely Christian God, is that a realization that itself realizes the most comprehensive Godhead ever known or apprehended, and so comprehensive that it is inseparable from every possible realization of totality itself? Could it possibly be that this is the very point at which Hegelian thinking is most truly and most absolutely offensive, most profoundly opposed to every other thinking, and yet here enacting a profound resolution and perhaps the only resolution of the totality of thinking?

Tom

Letter 39

To John D. Caputo (September 29, 2007)

Jack,

I am truly delighted to learn of your recovery, although I had heard rumors of it, and I gather that you are moving full steam ahead. Nor could I agree more with what you say about prayer and have even become convinced that our given or orthodox theologies are simply incapable of speaking of prayer, as is above all true of Barth and the Barthians and perhaps the Thomists as well. While music may well be our purest expression of prayer, poetry is our most articulate one, and I have long been persuaded that it is impossible to read or hear our purest poetry apart from prayer, and much the same could be said of painting, and in fact has often been said by great painters. The real problem to me is not the actuality of prayer but rather its real or actual meaning or identity. And this problem is most overwhelming to me in philosophy itself, and above all in our deepest or greatest philosophy, and here I regard the primary problem as Hegel. You may be unaware that nothing more unites theologians of all kinds than profound hostility towards Hegel, and yet Hegel alone among philosophers, as Findlay points out, transformed the deepest Christian dogma into purely philosophical categories, as above all manifest in an absolute kenosis or self-emptying.

No theologian has ever been able to think that kenosis, or purely think it, but this is the very essence of Hegel's thinking, and it can be understood not simply as a transformation but as a transposition of prayer itself. Of course, it is very difficult to think of Hegel himself as a man of prayer, but he never ceased to be a theologian, and even if a knowledge of theology is disillusioning about the prayer life of theologians, we cannot escape the realization of the distance between thinking and praxis.

So it is that I have long known the Faustian identity of real thinking, or of a truly modern thinking, one captured in Cornelio Fabro's marvelous history of modern philosophical atheism, *God in Exile*, but far more purely captured throughout the history of modern literature. Yes, Hegel is a Faustian thinker, even a Satanic thinker, as witness his profound atheism, but that atheism is inseparable from a purely kenotic thinking, and even makes possible a purely kenotic thinking, so that at last it is possible to understand not only the Incarnation but the Creation itself as an absolute self-emptying or self-negation. Is that a Satanic understanding? Many would have it so, but it is also possible to understand it as a passage of the depths of prayer into a purely conceptual thinking, a passage perhaps occurring in Plato and Neoplatonism, but not occurring in modern philosophy until Schelling and Hegel, unless it already occurs in Spinoza (which I must confess I believe is true). So then is it possible to say that a deeper thinking, or a deeper thinking for us, is simply impossible apart from prayer, and can be understood not as a transformation of prayer but rather as a transposition of prayer? Is this why a deeper thinking is inevitably a meditative thinking, one impossible apart from a genuine contemplation, a contemplation engaging our deepest depths, and even realizing those depths, even if such realization is a negative or Faustian realization, or one inseparable from damnation itself. Here, I tread upon my own theological way, but I am not alone in associating a deeper thinking with a genuine self-negation or self-emptying, and so, too, a deeper thinking is inevitably a solitary thinking, but a genuinely solitary thinking is inseparable from prayer.

Now that I am over eighty I am newly free, and free if only because I am at last free of the academic world, and I think that nothing has been more destructive of philosophy than its academic embodiment, one that for the first time has wholly dissociated philosophy from prayer or meditation. So, too, is academic training truly self-destructive at this crucial point, as perhaps most manifest in literary studies, literary studies dissolving literature itself, just as academic philosophical training so often dissolves philosophy itself. One of the reasons that I ceased engaging in documentation in my writing was an attempt to realize a meditative writing, of course, this did not occur until after I gained tenure and was safe, but it nonetheless was an attempt to realize what I knew was my vocation, and one initially attempted in a failed attempt to become a contemplative monk. Nor did I dissociate a monastic vocation from a theological vocation; this was rather my first real attempt at realizing a theological vocation, and a theological vocation inseparable from prayer, even if it is distant from all that is commonly manifest as prayer. As I mention in my memoir, I think that prayer is our deepest theological problem,[1] just as I also think that it is precisely the death of God that releases the deepest or the purest prayer, so that it is the Crucifixion that is the real center of a uniquely Christian prayer, just as it is the breaking or the death of God in the Eucharist that releases the purest communal prayer. And is not our Eucharistic theology our weakest theology because it most demands, or most overtly demands, a theology of prayer?

Jack, for one reason or another, I have been conversing about you recently, and finding that I disagree with what some regard as your real work. Without going into this, let me simply say that I have found you to be genuinely theological throughout your work, and from me that is not a complaint but a compliment, just as I rejoiced in your Syracuse appointment. So I have never doubted that you are a man of prayer, but now this can become open, or rather can be more openly manifest. Please forgive me if I share this communication with a few theologians, but we seldom communicate about prayer, and I am more guilty of this than most.

Tom

Letter 40

To Edward S. Casey* (November 23, 2007)

Dear Ed,

I find myself dumbfounded by your extraordinary response to my little ethics chapter[1]; there is so much brilliance in your remarks as contrasted with what you are responding to, and not only brilliance but even all too concrete and illuminating suggestions for revision. Frankly I am deeply embarrassed that you gave such energy to this, you surely have far better things to do with your time, and you should not induce me to think that such writing of mine as this deserves this consideration. And I must confess that I did not realize that you are such an able editor, but I should have known this if only because you must be your own primary editor, and your all too substantial but nonetheless truly clear and forceful writing demonstrates the highest editorial standards, standards that I fear are alien to me. Thus even though I will attempt it I do not think that I am capable of writing a persuasive account of our ethical crisis today, I simply assume that the reader knows this all too well, which perhaps is an illusory assumption. Yet there is another ground here, which I should state. At the age of 80 I have now virtually lost all interest in attempting to persuade others, at bottom I am now attempting to address my own crisis, or my damnation if you prefer, and to do this in the face of ultimate and immediate judgment. So let me attempt briefly to state what I am attempting in this ethics project, for simply stating it to you will be illuminating for me.

You may be unaware that my earliest writing for publication, which goes back to 1957 (50 years!), is grounded in an apprehension of the ultimacy of the Nothing, and whether directly or indirectly all of my writing ever since is an attempted mediation between our nihilism and an apocalyptic faith or enactment, this is what grounded my attempted recovery of the Christian epic tradition, and this has been at bottom an attempt to discover a truly if not absolutely new theological language. Originally I wanted to be a Greek scholar, but a realization of my all too limited philological talent soon induced me to abandon this, although this could have been a factor in my immersion in New Testament scholarship, a scholarship that when I was exploring it had a profound theological ground, and a theological ground now apparently absent from all biblical scholarship. The New Testament scholars who must affected me were Bultmann and Schweitzer, commonly apprehended to be virtually opposite theologically, but I learned through his sister that Bultmann was the only German New Testament scholar who responded positively to Schweitzer's great book on Paul,[2] and I found that in writing my chapter on Paul in my book on Jesus[3] (a book dedicated to Schweitzer), that it is possible to effect a kind of *coincidentia oppositorum* between Bultmann and Schweitzer, or between an existential and an apocalyptic understanding of the New Testament. I have become fascinated that that before New Testament scholarship discovered the apocalyptic Paul, Heidegger could not only apprehend the apocalyptic Paul, but in my judgment this apprehension is a fundamental ground of *Being and Time*, and most concretely so in its understanding of death, a death that is ultimately an apocalyptic death.

I have long been persuaded that Schweitzer is our greatest interpreter of New Testament ethics, and that this is not unrelated to his own sanctity, a sanctity unique among major biblical scholars, and just as Schweitzer was the first to establish a full continuity between the ethics of Jesus and the ethics of Paul, it is this very demonstration that most fully draws forth the apocalyptic ethics of the New Testament. One reason for my disgust with the world of Christian ethics is that it is so alienated from an apocalyptic ethics, and thus alienated from the New Testament itself, but this alienation is comprehensive throughout the Christian theological world, a situation so absurd that it is possible to know my work as our only apocalyptic theology. Incidentally I am also distressed that all of the recent philosophical studies of Paul simply ignore the truly critical scholarly work on Paul, perhaps because in France and Italy theology is in such deep eclipse, but also I sense because these thinkers, too, are wholly alienated from an apocalyptic ground. If nothing else I think that my work on Hegel and Nietzsche has called forth a deeply apocalyptic ground of each of these ultimate thinkers, a ground truly essential to such revolutionary thinking, and one that has a genuinely parallel expression in the revolutionary epic enactments of Blake and Joyce. Did you know that Norman O. Brown responded to my book on epic[4] with a joyous astonishment that *Finnegans Wake* had finally been understood theologically? And this is not only an apocalyptic understanding, but a nihilistic one as well, or one only made possible by a uniquely modern nihilism.

I think that you know that one of the ultimate books for me has been Nietzsche's *The Antichrist*, and I am convinced that this book is not only a genuine and revolutionary recovery of the ethics of Jesus, but that this ethics for us can only be a nihilistic ethics, or an ethics only possible in a world or totality of apocalyptic ending. Here, we can see all too decisively how nihilism and apocalypticism are so integrally related to each other, and while Nietzsche seemingly confined his deepest investigations of nihilism to his notebooks, I think it possible to say that all of his real or original thinking is a nihilistic thinking, and yet nonetheless it is our deepest ethical thinking since Spinoza. And if only through Spinoza and Nietzsche we can know that it is only through a revolutionary movement that we can actually recover a genuinely ethical thinking, and one at a virtually infinite distance from everything that is manifest as ethical thinking in our world. Yet note the all too important point that for both Nietzsche and Spinoza every distinction between the indicative and the imperative has wholly disappeared, only this disappearance makes possible an actual ethical praxis, but this disappearance most clearly occurs in an apocalyptic praxis, or in that Kingdom of God in which God is all in all. Hence the call to perfection in the Sermon on the Mount is not a categorical imperative, or even an imperative at all, it is only possible as a consequence of the final dawning of the Kingdom of God, a dawning giving us that "perfection" for which Jesus calls. So it is that here the indicative and the imperative are indistinguishable from each other, and I intend to establish that this is the deepest ground of a genuinely apocalyptic ethics, an ethics ultimately enacted by both Jesus and Paul.

Yet this is an ethics also ultimately enacted by Blake and Nietzsche, an ethics in which an absolute Yes-saying is indistinguishable from an absolute No-saying, one in which "Self-Annihilation" is an absolute and total reversal, a reversal occurring in that apocalypse which is all in all, or in that Eternal Recurrence that is an absolute Yes and No at once. Now if our common apprehensions are our most illusory apprehensions, this is illustrated by the common apprehension that the God of the Old Testament is the God of Judgment and the God of the New Testament is the God of Compassion, for damnation and Hell are wholly missing from the Old Testament, whereas they dominate the New Testament, including both Jesus and Paul. For only an absolute No makes possible an absolute Yes, only an absolute judgment, a judgment enacted by both Jesus and Paul, makes possible the "eternal life" of an absolute Yes, and just as for both Paul and the Fourth Gospel Crucifixion *is* Resurrection, that is the absolute No-saying that *is* absolute Yes-saying, and one only made possible by apocalypse itself. Accordingly, the absolute Yes of apocalypse is indistinguishable from the absolute No of apocalyptic judgment, as so purely enacted by the most overtly apocalyptic book of the New Testament, the Book of Revelation. No biblical book is more missing from the world of modern theology than is the Book of Revelation, but the Book of Revelation had a profound impact upon both Blake and Joyce, and if only though Joyce and Blake we can know the language of the Book of Revelation as a truly actual language, and one embodying an ultimate and final enactment.

Now even if such an enactment is wholly missing from the world or worlds of ethics, it is at the very center of an apocalyptic ethics, which is perhaps why an apocalyptic ethics is so invisible in our ethical discourse and our ethical worlds, but therewith is likewise missing an absolute or total Yes-saying, as so deeply enacted by Joyce, Nietzsche, and Blake. Yet that Yes-saying can be understood as a genuine renewal of a biblical Yes-saying, one most overtly present in the apocalyptic language of the New Testament, a language and enactment only made possible by the prophetic revolution of Israel, a revolution in which apocalypticism is truly born. How fascinating that both philosophy and theology now appear to be closed to the prophetic revolution, and hence closed to an absolute No-saying that is an absolute Yes-saying, or an absolute judgment that is an absolute transfiguration, as first fully and wholly actual in the Book of Isaiah, a book whose transformations in the course of the Exile made this the most apocalyptic book in the Hebrew Bible. This is also the biblical book that apparently most deeply affected Jesus, and if this is the book most fully repeated or renewed in the synoptic gospels, that is a repetition in the Kierkegaardian sense, for it is a forward moving repetition, and precisely as such an apocalyptic repetition, and one occurring in the actuality of immediacy itself. But this is an immediacy that is wholly lost in our common ethical discourse, a discourse that if only at this point is truly the opposite of apocalyptic discourse, and only an absolute reversal of our common ethical language can open us to a pure immediacy, one fully effected by both Nietzsche and Blake.

Ed, I fear that I must rest at this point, and even if rest is impossible for this old man, my restlessness drives me theologically, and so I hope that it never ends.

Tom

Letter 41

To Friends (November 28, 2007)

Friends,

I just read or mused through a cover story in *Time* magazine on recent scientific studies of good and evil[1] that is the most trivial such story that I have ever encountered, and there is no mention here of the way in which evolutionary biology has been so centered on the evolution of altruism. But this is the subject of *The Altruism Equation: Seven Scientists Search for the Origins of Goodness* by Lee Alan Dugatkin (Princeton: Princeton University Press, 2006). This book traces the history of the scientific debate about the origin of goodness from Darwin to the present through an extraordinary cast of characters, including Pyotr Kropotkin, although its real hero is the English mathematical biologist W. D. Hamilton.[2] Hamilton's Rule, which states that relatives are worth helping in direct proportion to their blood relatedness is here claimed to be as fundamental to evolutionary biology as Newton's laws of motion are to physics. For it is Hamilton who has perhaps most affected evolutionary biology, as is shown here through his ultimate impact upon such evolutionary biologists as Richard Dawkins[3] and E. O. Wilson[4], and the debate about evolution here largely revolves about the debate over the evolution of altruism.

Can you believe that at a crucial point in this development, and at one of its centers, the Biological Sciences Division at the University of Chicago, there was a firm belief that it was the biologist's job to "discover nature's moral prescriptions and thereby serve as a savior." These are the words of the then chairman of its Biology Department, Warder Allee[5], one of the pioneers in the field, and they apparently could characterize the aim of most of the major evolutionary biologists. While some of these such as Dawkins are passionate atheists, one of the most creative of them, George Price[6], was originally an atheist and was then converted, believing that God had inspired his biological work, then becoming an altruistic saint who committed suicide while in poverty. The center here is Hamilton's Rule which consists of the following: If a gene for altruism is to evolve, then the cost of altruism must somehow be balanced by compensating benefits for the altruist, because relatives may carry the gene for altruism as well. Phrased in the cold language of natural selection, Hamilton's Rule recognizes that a gene for altruism can spread if it helps copies of itself residing in blood kin, and thus the engine of goodness lies in the family unit. Here, it is even possible to make general predictions about the evolution of altruism, for the more related individuals are genetically, the more likely altruism is to evolve. And apparently it was the intense work on the evolution of altruism that established the deepest scientific ground of evolutionary biology.

Moreover, there is a parallel to this evolution in the evolution of "spite," the word here employed—although I would employ the word evil. Only rarely if at all has spite or evil been recorded in animals, but there can be demonstrated in humanity an evolution of spiteful behavior, a behavior commonly directed toward those who are very unlikely to have the same genetic composition. Yet it is not spite or evil that has gathered the attention of evolutionary biologists, it is altruism or even goodness itself, as a strange if not inverted Platonism would seem to dominate modern biology. Hamilton has been hailed as the most distinguished Darwinian since Darwin, and if evolutionary biology is now perhaps the most influential of all of the sciences, it is truly strange how in this perspective it can appear as a genuine expression of theology.

A puzzled,
Tom

Letter 42

To Walter A. Strauss (December 4, 2007)

Dear Walter,

I now write you while you are again in a painful condition, one beyond any I have known, and in my own way I am attempting to cheer you up. A short time ago I wrote a note on science and altruism, which I enclose[1]; there has been virtually no response to this, and I wonder if people are aware of how extraordinarily important it would be if the chasm were closed between Science and Spirit or science and morality. While the book that I discuss here is not important in itself, its thesis certainly is, and that is that modern evolutionary biology has realized its deepest ground in research on the evolution of the altruism gene, a research whose role in evolutionary biology is fully comparable to the role of Newtonian mechanics in modern physics. Many of the scientists who led this research are specialists in insect behavior, and here we can see animal altruism in its clearest form, as witness an ant colony. Indeed, W. D. Hamilton, who did the most crucial and influential mathematical research in evolutionary biology, and whose own research was done in Brazil on giant beetles, declared in a late published paper that he wished to be buried with his beloved beetles, so that "they will enter, will bury, will live on my flesh; and in the shape of their children and mine, I will escape death."[2]

While it may be difficult to accept the possibility that animal altruism is a genuine and even pure expression of goodness, and above all so in a world in which goodness is virtually unspeakable, evolutionary biologists have apparently been deeply engaged in a truly disciplined search for the origin and evolution of goodness, and one that has established the deepest ground of evolutionary biology itself. How odd that it is mathematical biology that is most powerful here, one presumably more powerful in evolutionary biology than in any other biological science, and one that has most decisively realized itself in the quest for understanding the evolution of the altruism gene. An altruism gene? And an altruism gene that is an absolutely decisive factor in evolution itself? Could it be that evolution itself *is* the evolution of goodness? A strange and even scientific Hegelianism? I find such questions irresistible here, and wonder why others do not. Are we simply forbidden even to raise the possibility that the purest science could be an expression of an ultimate quest for goodness itself, and for the purest and most absolute goodness? While many are fascinated by altruism, virtually everyone believes that at most it is an innocent illusion, and yet it could play such a role in this truly important and overwhelmingly influential scientific research!

The hero of *The Altruism Equation* is W. D. Hamilton, perhaps the most mathematical of all biologists, thereby being the most abstract thinker in this realm, and yet a thinker wholly given to the quest for the origin and evolution of goodness itself. Yet is this so absurd or so fantastic? Could we not say much the same of Hegel, and in a purely dialectical sense of Nietzsche, too, to say nothing of visionaries such as Dante, Milton, Blake, and Joyce? The very mention of these primal names evokes my own quest for the origin and evolution of the Christian epic, a quest that is seemingly of little or no interest to anyone, and I have never been able to understand this. Not even in the theological world has this quest evoked a response; it is as though it is simply nonsense, or Altizer's mad dream! Nor has my attempt to understand epic itself had any impact; epic as the only genre or mode that is a fully historical or realistic mode, and in its Christian expression a genre that is interior and exterior or psychological and cosmic simultaneously. It is a genre also embodying our most radically ethical expression, and a radically ethical expression that is inevitably our deepest heterodoxy. How can it be that this is of so little interest to others?

All this might well be yet another way of preparation for death, or an acknowledgment that an ultimate and final death has already occurred. I have often thought that this motif is embodied in all the higher and purer expressions of mythology and religion, and perhaps nowhere else may we discover a more genuine ecumenical or universal theology.

Hopefully in spite of everything,
Tom

Letter 43

To Friends (October 30, 2008)

Friends,

I am now reading Richard Kearney's *The God Who May Be*,[1] and while I very much admire this book and its deeply scholarly author, I am deeply disconcerted by its mythological language, never before having encountered this in a serious contemporary philosopher, or, indeed in any modern philosopher at all. Kearney is both Irish and Catholic, but he is certainly not a Joycean, and this does pose the question of whether or not it is actually possible for a contemporary Catholic to be non-mythological, for while this is certainly true of Leahy,[2] Leahy has yet to have an impact upon the Catholic world. One of the fascinating things about *The God Who May Be* is that it employs so much biblical and Patristic citation and yet remains resolutely free of all critical biblical or historical scholarship, it is almost as though it were written in a pre-modern world. Yet is Catholicism itself by necessity either pre-modern or postmodern, as manifest most clearly in its bondage to the mythological? Few realize how overwhelmingly important demythologizing has been in modern theology, even deeply affecting Barth himself despite his profound opposition to Bultmann, for one simply cannot discover a genuinely mythical language in Barth, and perhaps at no other point is Barth's theological genius so manifest. I think that there is a genuine parallel to this in Rahner[3] although not in Lonergan,[4] and this is perhaps a crucial dividing line in modern Catholic theology, although one that for obvious reasons cannot become fully open or manifest.

I think that demythologizing occurs most profoundly in Hegel, and while at this point Hegel was deeply influenced by Spinoza, in Hegel and in Hegel alone among philosophers the whole body of Christian dogma is transformed into a purely conceptual thinking, as most purely expressed in a uniquely Hegelian self-negation and self-emptying, a genuinely and profoundly kenotic thinking, and as such a philosophical realization of the Crucifixion itself. Hegel is the most profoundly Pauline of all philosophers, as most manifest at just this point, a Paul who had never been previously realized philosophically, and a Paul whose rebirth in Luther effected an ultimate revolution, one not only renewed but universalized in Hegel. Thus Hegel is the first philosopher of the Crucified God, or of the death of God, a death of God now realized in a purely kenotic thinking, and the purely kenotic thinking of an absolute self-negation and self-emptying. Paul knew the Crucifixion as apocalypse itself, an apocalypse truly reversed by the whole movement of a uniquely Christian theology, as the absolutely forward movement of apocalypse is transformed into the absolutely backward movement of eternal return, an eternal return not truly or actually reversed until the advent of modernity itself.

Accordingly, it is the seventeenth century and not the sixteenth century that is the truly revolutionary century, one realized not only in the scientific revolution but in a philosophical revolution that only realizes a decisive resolution in German Idealism, an idealism that turned the world of philosophy upside down. This is fully manifest in the reversers of German Idealism, Kierkegaard and Marx, but it is most purely and comprehensively embodied in the *Phenomenology of Spirit* and the *Science of Logic*, both of which effect an absolute demythologizing, and an absolute demythologizing of the deepest depths of the uniquely Christian God. Previously a realization of those depths had only occurred in an absolutely negative apophatic theology, a theology that all too significantly was always an orthodox theology, but a purely Hegelian thinking is the most heterodox thinking yet realized historically, and it absolutely transforms theology itself. Not even Augustine or Aquinas realized such a total transformation of theology, yet the Hegelian transformation of theology is inseparable from an absolutely new Bible, a Bible that is the consequence of a uniquely modern criticism and scholarship, a criticism itself effecting a profound demythologizing, and one wherein the sacred biblical text undergoes a total metamorphosis into a fully human text, as revelation itself undergoes an ultimately incarnational movement.

Nothing is more theologically revolutionary than modern biblical criticism and scholarship, which is either deeply resisted or purely suspended in all forms of both fundamentalism and neo-orthodoxy, and this ultimate fissure can even be traced to that Tubingen seminary where Hegel, Schelling, and Hölderlin were born, and born together as ultimate modern revolutionaries. Now just as all forms of political radicalism have been so deeply tamed or defanged in the contemporary world, this is likewise true of theological radicalism, and while biblical criticism has not come to an end, it has ceased to be theological, thereby most profoundly differing from its predecessors. One of the ironies of the twentieth century is that its most influential New Testament scholar, Bultmann, was the founder of the theological school of demythologizing, one that had a profound impact upon New Testament scholarship, even including Catholic New Testament scholarship, and yet the goal of this demythologizing was not the dissolution but the recovery of the deepest ground of the New Testament, one that in this perspective had been wholly lost by Christian theology. This is just the point at which Barth and Bultmann are so close, and while Barth never understood the demythologizing movement, he had earlier effected a genuine demythologizing in his revolutionary commentary on the Epistle to the Romans, which had a profound effect upon Bultmann. Bultmann himself was a great Pauline scholar, and just as his seminar on Paul had a deep effect upon Heidegger,[5] it is Bultmann's *Theology of the New Testament,* which is the deepest theological incorporation of Heidegger's thinking, as here Heidegger's early goal of a renewal of a lost primitive Christianity is truly effected.

It is also ironic that the recent philosophical renewals of Paul have ignored both Barth and Bultmann, who are the greatest modern exegetes of Paul, and who precisely thereby created a genuine theological demythologizing, and a demythologizing directed to an ultimate renewal of the Bible. This is what is so strange to the non-theological mind, which inevitably is a non-dialectical mind, or a mind that cannot realize that an ultimate theological negation can precisely thereby be an ultimate theological affirmation. This is fully manifest in both Barth and Bultmann, and just as nowhere can one find a greater assault upon Christianity than in Barth's *Epistle to the Romans*, nowhere else is there a greater theological affirmation of Christ, and full parallels to this dialectical theology may be discovered in the *Theology of the New Testament*, a theology that above all other theologies is an actual renewal of the New Testament. Of course, there is not a theology of the New Testament, there are multiple theologies of the New Testament, and each of these deeply differ from our theological traditions. This is a realization occurring throughout our New Testament scholarship, but perhaps nowhere so decisively as in the demythologizing movement, which once was a seemingly irresistible movement, even if subsequently it seemingly has vanished. Or has it vanished? Are not quests for a lost Jesus even now occurring, and quests for a revelation that actually speaks in our language, and speaks in the fullness and not the emptiness of our language? Surely a religion is comprehensively manifest now that is the very opposite of any possible genuine Word of revelation, a situation manifest in 1918 that called forth the birth of dialectical theology, and a dialectical theology that was the deepest biblical theology yet realized, and perhaps most biblical in demythologizing itself. Who today could imagine a world of biblical scholarship that is profoundly theological, and yet this was comprehensively manifest in the twentieth century, and even fully manifest in the most radical biblical scholarship. Perhaps there has never been a more radical biblical scholar than Bultmann, but there has also never been a more fully theological biblical scholar than Bultmann, and these dual dimensions of his work are simply inseparable. But this is also true of Schweitzer, and Bultmann was the only major New Testament scholar who positively responded to Schweitzer's great book on Paul when it was published in 1927, a book calling forth a purely apocalyptic Paul, and a purely apocalyptic Paul who is simultaneously the purely ethical Paul.

This, too, is demythologizing, one surely demythologizing historical Christianity, an historical Christianity that was initiated by its own ultimate transformation of an originally apocalyptic Christianity, and just as Bultmann, too, understood the original or the historical Jesus as an apocalyptic Jesus, the demythologizing movement profoundly understood that only an ultimate transformation of Christianity could be a genuine renewal of Christianity. Not even Kierkegaard could understand this, even if Hegel did understand it, but that understanding initiated a genuinely modern and a genuinely radical theology, and a truly modern theology that could only be a radical theology. Nowhere is this more openly manifest than in demythologizing, and all too ironically demythologizing is directed to the recovery of the New Testament, and the recovery of a lost New Testament, and one most lost in the Church itself. This Kierkegaard, and above all the late Kierkegaard, did decisively understand, so that dialectical theology began as a Kirkegaardian movement, as most manifest in Barth himself. While Barth was forced to renounce his original Kirkegaardian ground in creating the *Church Dogmatics*, demythologizing, too, is absent from this dogmatics, or present only in its most radical sections, for demythologizing is inherently radical, and most radical in its attempted recovery of a truly lost New Testament.

Tom

Letter 44

To Friends (April 10, 2009)

Beloved,

Many philosophers today are speaking of absolute gift, but theirs can only be a purely abstract discourse, and never a discourse evoking God, a God who is truly unspeakable today. Our world is a world in which Good Friday is everywhere and nowhere at once, for the death of God is nowhere and everywhere simultaneously, and if the realization of the death of God inaugurated a uniquely modern world, the consummation of that world has inaugurated a truly new godlessness, a godlessness occurring in a world in which there is seemingly a ubiquitous presence of God, a total presence occurring only by way of a total absence, and a total absence of that God who is the totally revealing or totally revealed God. The Christian knows that revelation as having most totally occurred in the Crucifixion, a crucifixion that is the absolute self-embodiment of God, as the totality of Godhead here undergoes an absolute self-emptying or self-negation, thus realizing a total kenosis, and a uniquely Christian kenosis. This is that absolute event that is celebrated on Good Friday, a celebration that is an anamnesis of absolute kenosis, and hence the renewal of an absolutely self-emptied Godhead, or of absolute emptiness itself.

White is the liturgical color of Good Friday, a whiteness most ecstatically celebrated in our great American epic, *Moby-Dick*, an epic that can be understood as enacting a cosmic Good Friday, a cosmic enactment of the kenosis of God. Yet, here, the kenosis of God enacts the absolutely Alien God, a God who is truly a *horror religiosus*, a *mysterium tremendum* wholly emptied of all *mysterium fascinans*, as a crucifixion is realized emptying Godhead itself of its original majesty and glory, and thus emptying the very substance or body of all pre-incarnate Godhead. Nothing less than such an emptying can truly be known as the death of God, or can actually be celebrated on Good Friday, thus the anamnesis of Good Friday is a renewal of this emptying, a renewal here and now of absolute kenosis. Only such a kenosis can truly be known as absolute gift, as that absolute grace that is absolute kenosis or absolute death, the sacrifice of Godhead itself, a sacrifice embodying the absolute gift of the Godhead, a sacrifice that alone is the actuality of absolute gift. Inevitably philosophical language is alien to this sacrifice, or all non-Hegelian philosophical language, and truly Hegelian philosophical language is inaugurated through the first philosophical realization of the death of God. Here, is a genuine philosophy of absolute kenosis, and our only one apart from Nietzsche, for it is Hegel and Nietzsche who most profoundly enact the death of God, hence Nietzsche and Hegel are our primal philosophers of Good Friday.

Hegel could know the non-kenotic God as Abstract Spirit or the Bad Infinite, even as Nietzsche could know it as absolute No-saying and absolute No-saying alone, an absolutely alien God known in the Christian world alone, and known in the horizon of the Crucifixion. Consequently, Good Friday is an *anamnesis* of absolute horror, the absolute horror of crucifixion, yes, but a gracious and triumphant horror, the horror of that absolute sacrifice that is absolute compassion. But such compassion is impossible apart from the realization of the absolutely alien God, or that dead body of God that is the consequence of crucifixion, a dead body that is an absolutely empty or alien body, and an alien body inaugurating a truly new *Angst*, an *Angst* known in the Christian world alone, and one the consequence of a truly new encounter with the Nothing. Indeed, an actual Nothing is a consequence of crucifixion, one not known philosophically until Schelling and Hegel, and one fully envisioned imaginatively in the uniquely Christian epic, and most fully envisioned in the revolutionary vision of Blake.

In Blake's mature vision, Good Friday is all in all, yet crucifixion *is* resurrection, just as resurrection *is* crucifixion, thereby effecting not only a renewal of Paul and the Fourth Gospel, but a renewal of that Christ who is the Crucified God. Moreover, this vision occurs in a world realizing the death of God, a death first imaginatively envisioned by Blake, and envisioned as inaugurating an absolute apocalypse, an apocalypse impossible apart from a realization of the death of God, just as Easter is impossible apart from Good Friday. Now even as the twentieth century can be known as realizing our most total and our most terrifying horrors, it can also be known as our most revolutionary century, and revolutionary in every domain, or every domain except for the religious one, a domain in that holiness itself now seemingly perishes. It is the pope himself who can now know our world as a truly godless world, a godless world that is a nihilistic world, and necessarily a nihilistic world in being a godless world, and if this is the first nihilistic world in history, it is the first world that is a consequence of an historical realization of the death of God. So it is that Good Friday is a decisive way of entering our world, of encountering that Nothing that is a decisive key to our world, and if Good Friday is a renewal of the Crucifixion, it is a renewal of that Nothing that is a consequence of crucifixion, and one apart from which resurrection is wholly illusory.

Good Friday is our most solemn day of meditation, a meditation induced by a renewal of the death of God, or a renewal of the most absolute grace that we have received, one transcending even the grace of creation, and transcending it because this is the grace that is the self-embodiment of God, and the only grace that is an actual self-embodiment of God. Inevitably we stammer in speaking of this grace, yet this is the grace that is realized in our greatest imaginative creations, and realized as that absolute death that is absolute apocalypse, or that absolute death that *is* absolute life. All too significantly Christian art was unable to envision the Crucifixion until almost a millennium of evolution, but then almost immediately the Crucifixion becomes the primary Christian ikon, and does so even more comprehensively in the late modern world. Now if it is death itself that induces the deepest awe, so, too, does it induce the deepest meditation, and not only the deepest but also the most universal meditation. While meditation upon death is universal in the history of religions, Christian meditation is nonetheless unique in being centered upon an actual death, and even an historical death, an historical death that both Hegel and Heidegger can know as the center of history, and the center of history because it is the one absolute event in history. So that meditation upon this death is inevitably an historical meditation, or a meditation upon the horrors of history, horrors that are fulfilled in the Crucifixion, a crucifixion alone effecting an atonement of those horrors. The cross is our most universal symbol, or most universal actual symbol, and it is a symbol of atonement, and of universal atonement, one that Blake envisioned as "Self-Annihilation," and the self-annihilation of the "Universal Humanity Divine." Therefore the Good Friday meditation is a universal meditation, one enacted wherever there is an actual meditation upon death, or an actual meditation upon the ultimacy of death.

Yet such meditation is truly a liberating meditation, one liberating us from our deepest bondage, or our deepest interior bondage, an interior bondage of innumerable chains, but chains that are challenged or dislodged by meditation itself, and above all by meditation not only upon death but upon an absolute death. To meditate upon that death is to meditate upon the actuality of an absolute compassion, a compassion only realized through an absolute death, and through the absolute self-emptying or self-negation of that death, that is the self-negation or self-emptying that is recalled or renewed in a Good Friday meditation. This recall or renewal is an *anamnesis* of the Crucifixion, an *anamnesis* of absolute death, and one mediating that death to our own center, a mediation that is necessarily an atonement, and an actual atonement of our very center. Sacrifice can be understood as the most ultimate movement occurring throughout the history of religions, nothing else is more truly universal in the history of religions, and if Christianity is ultimately centered upon the Crucifixion, it is centered upon a death of God that is truly unique in the history of religions. Nonetheless a meditation upon the death of God is in genuine continuity with the world or worlds of religion, here we do truly enter a universal humanity, or the "Here Comes Everybody" of *Finnegans Wake,* and just as the *Wake* continually enacts a universal mass or sacrifice, the text of the *Wake* is our most powerful modern liturgical text, and one enacting a truly universal liturgy.

If only in this perspective we can understand that a Good Friday meditation is a truly liturgical meditation, one occurring in a liturgical body or world, a truly corporate world, and thus a truly bodily world. Paradoxical as it may appear, meditation itself can and does realize a bodily world, and above all so a meditation upon absolute death, which is a meditation upon an absolute "Self-Annihilation," a "Self-Annihilation" that Blake envisioned as the ultimate source of Body or Energy. This is the Body or Energy that a Good Friday meditation can call forth, and precisely by its *anamnesis* of an absolute compassion, an absolute compassion only realized by the actuality of an absolute death. So it is that in meditation upon this death we do call forth a new energy and a new life, a new life that is not our own, but is the consequence of an absolute gift, and the absolute gift of an absolute compassion. That is the compassion we know in a Good Friday meditation, one that is given us in that meditation, a gift that is an ultimate liberation, or is the actual promise of an ultimate liberation. This is the promise that is enacted in a Good Friday meditation, and enacted in the very taste of death itself, or in the taste of an absolute death, that one death that is the ultimate source of liberation.

Tom

Letter 45

To Friends (May 8, 2009)

Friends,

Pardon me once again as I employ you as a sounding board for a problem of my own but I sense that this time I am speaking of a problem that is uniquely my own. Some of you will know that I have embarked upon a project that Lissa urged me to do, a study of the genealogy of nihilism, a genealogy that has been an underground presence throughout my work. Its first chapter is entitled "The Advent of the Nothing," which came somewhat easily until I entered the heart of it, and that is the role of ritual in this advent. Now ritual has been a fundamental challenge to me throughout my work, although I have done very little on it, and when younger I was affected by the myth and ritual school of literary criticism that insisted that myth is wholly unreal apart from ritual, just as the origin of myth is inseparable from the origin of ritual. A comparable approach is present in the Durkheimian tradition, and it has always been a challenge of mine to relate this tradition to the Weberian tradition, Weber and Durkheim are the two primal figures of sociological theory, and having been trained by Wach[1] I was molded in a sociology of religion shaped by these figures, and one that has never been integrated into theology. So, too, the Nothing has been touched upon but never integrated into theology, indeed, I believe that there is a deep prohibition here, and one present in philosophy as well as theology, my favorite example being the refusal of the philosophical world to open itself to the deep origin of Heidegger's *Ereignis* in the final redemption of Faust through an ultimate union with the Nothing in *Faust II*. This ultimate movement is I believe purely paralleled in the apocalyptic redemption of Satan in *Milton* and *Jerusalem*, although the one book on Goethe and Blake misses this crucial parallel.[2]

Now I must confess that as a good Nietzschean (and good Freudian?), I have always been attracted to the forbidden, and most attracted to the absolutely forbidden. Of course, the Nothing has become absolutely forbidden inasmuch as it has been evoked, and this deeply occurs in late modernity, here there is a unique epiphany of the Nothing, and a uniquely modern Nothing, a Nothing that as first unveiled by Kierkegaard is wholly *mysterium tremendum* and in no sense whatsoever *mysterium fascinans*. One has only to mention the names of Kafka and Beckett to make this clear, but if this is a unique advent of the Nothing, what is the relation of this advent to an absolutely primordial Nothing? This is where the deep problem of ritual now arises for me.

First I am attracted to ritual because despite its primal and ultimate importance virtually no real thinking has occurred about it, is this not real evidence that we are deeply if not absolutely forbidden to think about ritual, as witness the comprehensive weakness of our liturgical theologies, even including Aquinas. I once knew a Jesuit who had completed a dissertation on Calvin's Eucharistic theology, demonstrating that Calvin had purely and comprehensively dissolved every possibility of the Real Presence, and done so by demonstrating that Christ is actually in Heaven and in Heaven alone. Hence we can only know Christ by absolutely eliminating every possible sacramental act, and for a Weberian this frees us for an absolutely new autonomy, or an absolutely new praxis effecting a revolutionary transformation of the world. Note, too, Weber's demonstration that a uniquely Confucian ritual or *li* effects a radical secularization of the world, truly distancing the sacred, and thereby releasing a wholly new praxis that transforms the world.[3]

Let me return to the problem of ritual. I am attempting to understand a uniquely ritual power, one that can be discovered in ritual and in ritual alone, and one that is not only a human power, but one that is absolutely necessary both for a uniquely human power and a uniquely human community. Now as recent sociology has demonstrated, ritual occurs comprehensively throughout our world, we simply fail to understand it, just as we are profoundly alienated from any possible liturgical or ritual understanding. Now I am attempting to understand that our deep alienation from ritual is a consequence of ritual's integral relation to the Nothing, that ritual or pure ritual is an *anamnesis* of the Nothing, an *anamnesis* that we have never understood, just as we cannot understand *anamnesis* itself, and above all so not when we think of it as remembrance. Nothing is more primordial in our world or in any world than ritual itself, or pure ritual, a ritual than can be understood as being prior to myth, and prior to consciousness itself, or more primordial than any form of consciousness. Indeed, pure ritual is the purely primordial, but does not that inevitably associate it with the Nothing, a Nothing inseparable from the absolutely primordial, as manifest in every deep expression of mysticism? Yes, we can know the Nothing in a deeper mysticism, just as we also know the Nothing in the deeper expressions of the late modern imagination, but do we not also know or realize the Nothing in pure ritual itself, a ritual that is an enactment of the Nothing, and a redemptive or purgative enactment of the Nothing?

While these questions may appear to be odd, think of them in the context of *Finnegans Wake*, the supreme epic of our new world, and an epic whose text is the most liturgical literary text in the world, its very language, and scatological and blasphemous language, is a purely liturgical language, one continually enacting a *missa jubilaea*,[4] and an *anamnesis* that is the *anamnesis* of the death and resurrection of an absolutely primordial H. C. E. or Here Comes Everybody. Nowhere is a uniquely modern Nothing more comprehensive than in the *Wake*, one absolutely disrupting all language and all consciousness, but that disruption realizes an absolute chaos that is apocalypse itself or a New Jerusalem, an absolute Yes that is only possible as a consequence of the realization of the absolute No of an absolute Nothing. We are not simply called to an absolute purgation in the *Wake*, but a real reading of the *Wake*, which is inevitably an oral and communal reading, is a realization of such purgation, and as such it is pure ritual itself. Clearly such ritual is an *anamnesis* of the Nothing, and of a Nothing that is absolutely primordial and uniquely modern at once, or absolutely sacred and absolutely profane simultaneously.

Pure ritual as an *anamnesis* of the Nothing? Does not this make perfect sense in the context of *Finnegans Wake*, a context even drawing forth the meaning of *anamnesis* itself, one that we have never understood, and perhaps not understood because it is so deeply forbidden, and forbidden if only because it is a renewal of the Nothing. Certainly *anamnesis* is renewal rather than remembrance, here exposing the hollowness of all of our catechisms, but in pure ritual it is a renewal of that purely sacred that is purely profane, and this is just as true of *li* as it is of the Eucharist. Hence it is a renewal of "flesh," and not that *sarx* that Paul knew, or even the *soma* of the New Testament, but rather of a primordial Flesh or primordial Body. Could it be that every pure ritual is such a renewal, hence it is absolutely essential to the human as such, and to every human community or human body? Is pure ritual a pure power because it is such a renewal, more powerful in itself than any of our cultural or social expressions, and yet a pure power that as such is absolutely forbidden? And absolutely forbidden because it is a pure renewal of the Nothing?

Tom

Letter 46

To Friends (March 25, 2010)

Friends,

I am now reading Jacob Taubes's *Occidental Eschatology*,[1] which you will remember is a sacred work for many, and it is a truly unique work in its full and genuine integration of historical analysis and philosophical-theological thinking, moreover it is centered not in Judaism or the Hebrew Bible but rather in Christianity. It is also and even thereby an extraordinarily powerful study of apocalypticism, drawing forth apocalypticism in its unveiling of a deeper Christianity, and an apocalypticism that culminates in Hegel, Kierkegaard, and Nietzsche. This book was published in 1947, written when Taubes was 23 years old, and it is supposedly a doctoral dissertation although it lacks all signs of such. Clearly Taubes had then mastered a vast body of material, but he rethinks it so as to make it his own, and if you will remember the Taubes essay that I republished in *Toward a New Christianity*,[2] he can brilliantly deconstruct modern Christian theology by drawing forth its repressed apocalyptic origin and ground.

Let me give a sample of this book from its section entitled "The Life of Jesus": This reflects a serious study of New Testament scholarship, and also embodies a principle of Taubes that the beyond of history is the essence of history. A good deal of this section is given to an analysis of the Kingdom of God proclaimed and enacted by Jesus, and the *metanoia* that his disciples are to preach is not repentance but rather an act turning the world upside down, for the preaching of Jesus is revolutionary because it demands decisive action for the Kingdom of God. He demands the exodus of the people into the wilderness, only thereby can one escape the domination of this world, and it is in the wilderness that there will occur the redemptive event of the coming of the Son of Man. Yet this event fails to occur, and this disappointment is central to the life of Jesus, and if the whole history of Christendom is determined by the delayed Second Coming, then the first date in Christian history can be taken to be the nonfulfillment of this prophecy of Jesus. This nonoccurring event marks the decisive otherwise inexplicable turn of events in the work of Jesus, only now does Jesus know and reveal that the Kingdom can only be brought about by the passion of the Son of Man, and it is the suffering of the Son of Man that is the eschatological event of *kairos:* "In the Kyrios Christos cult the fate of a man becomes the symbolic and metaphysical center of creation" (p. 58).

This section is followed by one on Paul, which is far superior to the mediocre book on Paul[3] written by the disciples of Taubes after his death (Jewish disciples who knew little of Paul), and one can never know Taubes through such a book. I wonder what kind of response *Occidental Eschatology* will receive, and even hope that someone will ask me to write an essay review of it. In any case we are here given an important theological work and I hope that it will not be ignored by the philosophical world.

Tom

Letter 47

To Friends (March 31, 2010)

Friends,

I am continuing to deeply enjoy my reading of Jacob Taubes's *Occidental Eschatology*, and even reading it as a work of Christian apocalyptic theology, for this is not simply an exposition of Western eschatology, but a reenactment of it that is simply not possible apart from theological thinking. Taubes much affected me when I was a younger theologian, and I did meet him shortly before his death when he was in a pathological condition, this was through Edith Wyschogrod, and she and her husband Michael were deep participants in the Taubes circle. That was a Jewish circle, and Taubes was deeply Jewish, but he shared with Edith a deep participation in Christianity, and this is fully manifest in his book. And the book is truly remarkable as an apocalyptic enactment, I would even speak of it as an *anamnesis* or renewal of the historical realization of the Christian apocalypse, and even if its section on Jesus is perhaps its weakest section, it is the apocalypse of Christ that is celebrated in this book, and here philosophical and theological thinking are wholly conjoined.

What I am finding to be most enlightening is its renewal of Joachism,[1] a Joachism or revolutionary Spiritualism that is profoundly reborn in Hegel, and the *ecclesia spiritualis* or "eternal gospel" is here called forth as the deepest revolutionary ground of the West, and one whose collapse as a Christian eschatology is simultaneously its rebirth as a philosophical or transcendental eschatology. I have long attempted to think of myself as a theologian of the Spirit in the tradition of Joachim, which is also of course the tradition of Blake and Hegel, but just as philosophers deeply resist the theological ground of Hegel, they thereby resist its own most integral ground, even if Hegel himself knew little of Joachim or of Joachism. So far as I know Joachism was not even discovered historically until the twentieth century, despite the fact that it has been profoundly influential upon the modern world, and has even been pathologically reborn in twentieth-century totalitarianism. And it is intriguing that apparently the world of New Testament scholarship is wholly innocent of the "eternal gospel," just as is the world of theology, and here Taubes's book is truly important, and hopefully will have an impact upon these worlds. And could we be closed to the "eternal gospel" because it is so profoundly revolutionary, being as fully realized by Marx as by Hegel, and if Christian Marxists could become aware of this there could be a rebirth of Christian Marxism, which perhaps could only occur in a wholly non-Marxist world?

Yours hopefully,
Tom

Letter 48

To Friends (June 4, 2011)

Friends,

Yesterday I had a fleeing moment in which inspiration appeared to be fully at hand, one focused upon that which had always most baffled me, the integral relation between Athens and Jerusalem, and not the relation between reason and revelation, but rather the relation between an actual Greece and an actual Israel. Both, of course, are truly complex, but I intended to focus upon what Nietzsche knew as an Apollonian Greece and upon the Israel of the prophetic revolution. Both are truly revolutionary, as perhaps most intimately revealed by the role of death in each, a truly new and unique death, without any real parallel throughout the world. Many naively believe that both Greece and Israel shared a belief in the immortality of the soul, but this belief is alien to the great majority in both Greece and Israel, and above all alien to an Apollonian Greece and a prophetic Israel.

If Homer is the Scripture of Apollonian Greece, death here is truly unique, an heroic death impossible both for the gods and for the vast majority of men, and one only possible through *arete* or the purest virtue, as most fully embodied in Achilles, whose *arete* is inseparable from an early death. Now this is also a truly destined death, for heroic death is not only inseparable from destiny, but is a consequence of destiny, a destiny most decisive in the very actualization of death. Nothing is more integral to an Apollonian Greece than a destiny that is Destiny, an *ananke* that is an absolute necessity, but one truly other than every natural necessity. For the hero wills that death that is his destiny, a willing that can only be named in this destiny, for willing as such is alien to the ancient world, and was only discovered by Augustine.

Indeed, an heroic death fully parallels what Augustine knows as grace, but in no way whatsoever realizes the Beyond, or does so only in an instant, an instant that immediately vanishes, but that immediacy is a pure immediacy that is an incarnation of life or energy itself. Every beyond is absolutely alien to this immediacy, so, too, an Apollonian Greece is the very opposite of a Dionysian Greece, and Dionysian deity is the very opposite of Apollonian deity, an Apollonian deity absolutely unique to Greece, whereas Dionysian deity is universal. Homeric deity is the purest expression of Apollonian deity, a deity without any moral dimension whatsoever, and a deity appearing everywhere and nowhere at once, and if Apollo if the most Homeric of all the gods, he is also the most unique of all the Greek gods, and absolutely distant from Yahweh or the biblical Lord. This is a distance making possible an Apollonian grace realized in death itself, a death that is an heroic death realized wholly apart from either fear or *Angst*, and one thereby conquering both fear and *Angst*.

Of course, Homeric deity is wholly incapable of inducing *Angst*, just as its own movements are as gracious as any in the world, embodying a light that is wholly alien to darkness, and a light that can be a giver of a gracious death. Nothing could be further from this deity than Yahweh or the Lord, a Yahweh whose absolute judgment is an embodiment of an absolute death, and while Yahweh is the Creator, this is not manifest until the prophetic revolution, a revolution effecting the twilight of the gods, and even if this revolution is historically coincident with the Homeric revolution, no revolutions are more distant from each other than are these. And Yahweh assumes a whole new identity when apprehended from an Apollonian perspective, then He truly is the Alien God, the God of absolute judgment and absolute judgment alone, thereby the God of an absolute death, a unique death ending all life and grace, as for the first time eternal death is actually realized in the world.

While eternal death may well have been born in the Babylonian and the Iranian worlds, it does not become totally embodied until the prophetic revolution of Israel, only its absolute negation made possible an absolutely new faith, a new faith realizing a truly new death, a death that is invisible and unspeakable, thereby making possible a wholly new life. Such a life is the very opposite of any possible immortality, for it is a life given wholly here and now, a life free not of the actuality of death but of the consciousness of death, a consciousness of death that vanishes with the twilight of the gods.

Now that vanishing is actual both in an Apollonian Greece and a prophetic Israel, perhaps at no other point is each so integrally related to the other, at this point alone Apollonian deity is integrally related to Yahweh, as each makes possible an absolutely new affirmation, and an affirmation of earth and time found in no other historical worlds. But a new other-worldliness has overwhelmed each of these traditions or ways, as above all occurring in Christianity, a Christianity reversing its original ground by transforming an apocalyptic advent into an other-worldly quest, giving birth to a Greek or Eastern Christianity that is the most other-worldly of all traditions or ways, and an Eastern Christianity that created Christian dogma as a whole. This is the dogma that is most distant from an Apollonian way, and that could only have been created by reversing that way, thereby abolishing all life that is not a reflection of the Beyond, and all energy that is not a gift of the Beyond.

Now, and for the first time, immortality is all in all, a new immortality wholly alien to the pre-Christian world, and above all alien both to Israel and to Greece, only now can Plato be virtually deified, and an absolutely new Neoplatonism appear that can consume all thinking whatsoever. This is what Nietzsche knew as an absolute catastrophe, a catastrophe only reversed by an absolute realization of the death of God, a death that a new Zarathustra can know as absolute liberation. If only at this point there is a genuine continuity between Apollonian Greece and our contemporary world, each is only ultimately liberated by death, and by a free or willed death that is a destined or predestined death, a death of all eternal life that alone is the way to a fullness of life.

Tom

Letter 49

To Friends (July 18, 2011)

Friends,

I am temporarily obsessed with attempting to understand Max Weber's understanding of modern revolution, which so far as I know is very little apprehended or even attended to. Few go beyond his early book on the Protestant ethic but this is a fundamental foundation of his work and specifically so in so fully integrating the economic and religious domains, he is simply unsurpassed at this. Now Weber is one of our great historical thinkers, thus he created the category of a uniquely modern "innerworldly" asceticism, which arises at a unique moment of history, when for the first time there was an unprecedented inner isolation of the individual, one releasing a new and total damnation, removing the individual from all ancient religion and all institutions, and giving birth to a new and absolute responsibility. Then a purely and totally pragmatic action is first released in the world, one made possible by an absolute responsibility, and a responsibility inseparable from a new and total predestination, a predestination eternally willing our absolute responsibility, which is inseparable from our being eternally predestined to damnation. Even the redeemed are redeemed from that damnation, a damnation that is our deepest condition, and one impelling an absolute responsibility.

No one has understood predestination more deeply than Weber, or none since the creators of the doctrine of predestination, Paul and Augustine, and just as it was a radical renewal of Augustinianism that initially gave birth to modernity, it was a pure and radical expression of Augustinianism in Calvinism that released an absolutely new pragmatic power, thus creating what I am coming to think of as the secular revolution. Now the secular revolution is quite simply the modern revolution, a revolution comprehending the modern economic revolution, the modern political revolution, the modern social revolution, the modern scientific revolution, the modern imaginative revolution, the modern philosophical revolution, the modern religious revolution, and all of these revolutions are integrally related in being the secular revolution. This is what I think that Weber profoundly understood even if he didn't employ this language in speaking of it. But note that in each of these all too divergent spheres there occurs the revolutionary birth of a profoundly secular movement, one that had never existed before in any of these domains, but whose very advent issues in a total transformation, and a total transformation of the world itself. Yet this virtually apocalyptic transformation culminates in what Weber knew as the "iron cage," and the late Weber spoke of the fate of our age as a coming polar night of icy darkness and severity, and could even know this age of total reaction as having already closed in upon us.

Weber can be understood as being the first thinker to integrate Hegelian and Nietzschean thinking, and just as Weber is the most comprehensive of our great twentieth century scholars, he integrates all historical domains by a genuinely radical thinking, thus creating an absolutely unique sociological thinking. The center of this thinking is secularization itself, and here Weber is a deeply original thinker, for here secularization is profoundly religious and profoundly antireligious at once, it could even be understood to have been inaugurated by the prophetic revolutions of both Israel and China, and just as Weber is alone as a sociologist of the world as a whole, he is alone in his understanding of the revolutionary power of secularization.

Tom

Letter 50

To Friends (July 22, 2011)

Friends,

Yesterday in the midst of the terrible heat I watched my favorite DVD, "Ballets Russes,"[1] a celebration of the Ballets Russes[2] reenacting their history and giving us truly ecstatic moments of absolute ballet, and culminating with a reunion of their dancers when they were in their eighties and beyond. And how marvelous they are! Having seen their glorious dancing in far earlier moments we are nonetheless aware of a genuine continuity, and one bestowing upon them not simply a dignity but a grace that is seemingly all in all. But in this context, how can one look upon one's own old age? First, and this may well be simply an illusion, I sensed that I look younger than they did at that moment. This if true could be because I have lived a far less demanding life, or simply have a different genetic inheritance, or even have had a fulfillment in my life they did not in theirs. Remote as this last possibility is it nevertheless intrigues me, after all I do look upon ballet as our most sacred art, just as I regard ballet dancers and choreographers as our most sacred artists. How could I have had any fulfillment in my life going beyond theirs? Certainly it could not be in my life as such, but could it have been in my purest theological thinking when I was least myself and most given to my calling?

One reason that I venerate the ballet world is because of its absolute dedication to its calling, this came forth again and again in this film. For genuine sacrifice appears to be inseparable from genuine ballet, and seemingly far more universally so than in any other art or calling. Have you seen the film, *The Red Shoes*?[3] I deem it a great film, and its primal actor and choreographer and male dancer is actually Leonide Massine,[4] who is as great as any other figure in the historical world of ballet as demonstrated in the best book on the subject, *Apollo's Angels*.[5] Indeed, Massine must have been an old man when this film was made, but he is truly great just as is his choreography, and he may well be the greatest of all choreographers. I am eclipsed by Massine, of course, and here he even looks younger than I now do, but nonetheless I do not simply melt in any such comparison, or not in my own judgment. I bought *The Red Shoes* and have watched it several times, but I am most excited by Massine rather than by Moira Shearer[6] or Robert Helpmann,[7] and this may be a unique film in employing the actual one who is its primary creator or source, and it fully succeeds as comparable Chaplin films do not. And Massine looks younger here than I do now, and could that be because he even then embodied a primal power that I have lost? Yes, I must openly face this probability, and recognize that my aging has been less than it could have been, and perhaps even now is subject to the possibility of a genuine renewal.

Your aged,
Tom

Letter 51

To Friends (July 23, 2011)

Friends,

I have just finished rereading Max Weber's *Protestant Ethic*[1] for the first time in many years and I am stunned by it. The primacy with which it treats Predestination is amazing, and I would even recommend this book as perhaps the best way to enter this absolutely important but totally misunderstood doctrine. Now it is not just that all early Protestants believed in Predestination, but it was at the very center of their faith, and apart from understanding it you simply cannot understand Protestantism (which means that very few do understand it!). Now primal in Weber's understanding of Protestantism is his understanding of worldly asceticism or what he later termed innerworldly asceticism, this is only truly born with the advent of Protestantism, and it gradually revolutionizes the world as a whole. This occurs with the full birth of capitalism, which is inseparable from the actuality of modernity itself, and it is Weber who perhaps more than any other scholar reached a truly and fully integrated understanding of modernity, one comprehending its economic, its political, its social, its interior, its intellectual, and its religious domains.

Weber was only able to write *The Protestant Ethic* after having undergone a devastating mental breakdown, and this is the first of his truly major works, one making possible all that followed. Its thesis may be briefly stated: the religious valuation of restless, continuous, systematic work in a worldly calling, as the highest means of worldly asceticism, and at the same time the surest and most evident proof of rebirth and genuine faith, must have been the most powerful conceivable lever for the expansion of the spirit of capitalism. Calvinism rather than Weber's own Lutheranism is central in this book, and above all the Calvinistic doctrine of Predestination, a predestination occurring before the creation, and the elect are chosen only out of God's free grace and love, without any foresight of faith or good works, while the rest of humanity are denied God's grace as an exercise of God's absolute sovereignty and glorious justice.

The advent of this age was the advent of a new individualism and one bringing with it a truly new loneliness, for Weber maintains that the Father of the New Testament is now gone, His place taken by an absolutely transcendent God beyond all understanding, who with His wholly incomprehensible decrees has eternally decided the destiny of every individual and regulated the tiniest details of the cosmos from eternity: "God's grace is as impossible for those to whom He has granted to lose as it is unattainable for those to whom He has denied it."[2] Hence an overwhelming compulsion to discover whether or not one is among the elect, and this releases a new ascetic energy, driving one into a new praxis to actually test one's destiny, for the actual exercise of a vocation in the world can be a sign that one is among the elect. While I am critical of these formulations by Weber, I think that the calling forth of this problem is deeply important, and the new age of modernity truly does release a new and even absolute energy in the world, one can call it with Weber an innerworldly asceticism, but it does absolutely transform the world, and it does culminate in an "iron cage."

Your Weberian friend,
Tom

Letter 52

To Andrew W. Hass (July 29, 2011)

Dear Andrew,

Yesterday in writing you about Aristotle I evoked an itch in myself that has not yet dissolved, so I was led back to my first book *Oriental Mysticism and Biblical Eschatology* (1961), long since out of print and disappeared, and as I remark in my memoir although the book now embarrasses me so far as I know there is no other book on this important subject.[1] There are more pages on Aristotle here than I had remembered, and they revolve about a fundamental theological point, for Aristotle it is Becoming that is ultimate, and the very reality of Becoming necessitates the "withdrawal" of God. The book on Aristotle that most affected me was Werner Jaeger's,[2] a book that profoundly affected Heidegger, and the crux of Jaeger's thesis is that the decisive key to Aristotle is Aristotle's ever more progressive and ever more total movement away from Plato.

The first chapter of my book is on Greek religion, and despite its apparent irrelevance to the topic of the book, this deeply shaped the book as a whole, and did so by opening up a world that is so vastly distant from either biblical eschatology or Oriental mysticism. So I discuss Aristotle in the context of exploring Greek religion, a religion that is perhaps more unique than any other, as nowhere more manifest than in Aristotle himself, whose very understanding of God is absolutely areligious, and more areligious than any other understanding of God. God as the Unmoved Mover is perfect, eternal, and incorporeal, and God being unmoved is pure actuality (*energeia*), and the pure actuality of the cosmos as a whole. But God has no reality apart from this actuality, only thereby is actuality or Becoming ultimately real, and only thereby is science or philosophy actually possible. In this perspective, Aristotle's metaphysics can be understood as the opposite of Plato's metaphysics, just as Aristotle's metaphysics is progressively dissolved after the advent of Christianity, even if it is renewed in Jewish, Islamic, and Christian scholasticism. Yet the scholastic Aristotle is vastly distant from the historical Aristotle, perhaps infinitely distant, but an historical Aristotle that is renewed with every reversal of scholasticism, as most profoundly occurring in Hegel and Nietzsche.

I had not yet entered the world of Hegel when I wrote my first book, although I had been much affected by Jacob Taubes's Hegelian understanding of the secularization of modern theology, and it is fascinating how distant the secular Aristotle is from modern secularization. For the secular Aristotle is the teleological Aristotle, an Aristotle for whom an ultimate purpose or causality is everywhere, and even everywhere as the absolutely immanent God. Here, God is real precisely by being transcendentally unreal, a transcendental unreality that is inseparable from the actuality of the Aristotelian God, and if that actuality is the *energeia* of the cosmos as a whole, this is a uniquely Aristotelian *energeia*. Is a theological understanding of such an *energeia* possible? And has it ever occurred? Or is this an ultimate challenge now lying before us?

Your Aristotelian friend,
Tom

Letter 53

To Friends (August 2, 2011)

Friends,

I just read an excellent review article in the *New York Review of Books* on Alfred Kazin's journals[1] and was stunned to see the enormous emphasis it placed on Kazin and God. Kazin and I were colleagues in the English department at Stony Brook, and I had many conversations with him, the question of God never came up, and yet I suspect that I was the only theologian whom Alfred knew. He did deeply respect Hannah Arendt whose Gifford lectures demonstrate that she is a truly major theologian but I doubt if Alfred would have thought of her as such. In his memoir, Alfred presents himself as being deeply Jewish, but a heretical Jew, one wholly alienated from Rabbinic Judaism, and even in some sense alienated from the Hebraic God. Nonetheless God is all important for him, and here he deeply identifies himself as an American, and an American in the tradition of Emerson, Melville, and Stevens, one in which God is overwhelmingly important but only as the absolutely heterodox God.

Kazin was primarily devoted to American literature, and he is perhaps its best critic, but he was also a Blakean as this review emphasizes, although we never discussed Blake. Indeed, it would be my judgment on the basis of this review that Kazin's God was closest to the Blakean God, and while so far as I know I am the only theologian who is a Blakean, I could now greet Kazin as a Blakean, and even suspect that this is a primal theological way of encountering our deeper intellectuals today. Blake had an absolutely terrible image until he was resurrected by Yeats, but ever since he has had an absolutely positive image, and not even T. S. Eliot dared to attack him. Yet he is our most revolutionary prophet, and most revolutionary religious prophet, even forcing Nietzsche into a lesser light, but no one has dared yet to do a study of Blake and Nietzsche, as though this might create a conflagration beyond our ability to bear.

Come to think of it, *The New York Review of Books* has been weak on both Blake and Nietzsche, just as are most of our universities, perhaps this is forbidden ground, and if so that is just the ground that we must seek in searching for a site to bring together God and the contemporary intellectual. It fascinates me that Kazin was so silent about God throughout his life even while so deeply seeking God, is he thereby a paradigm for a great many others, and even others who are commonly identified as atheists? There has been much very foolish writing about atheism recently, yet this appears to be taken seriously, as though we had developed a literal void on the subject of atheism, despite the fact that atheism as Atheism is so powerful in full modernity. Our recent childish atheism seemingly does not even know Nietzsche, and perhaps such ignorance is truly characteristic of our world, but if so that leaves little possibility for genuine theological dialogue.

Nonetheless we must proceed as though there is some such possibility, and I am encouraged by the space devoted to God in the Kazin review, even if discouraged by the apparent fact that virtually all of our publications when occasionally discussing God can speak only of the orthodox God, as though the orthodox God is the only God, and this to me is the truly dangerous or blasphemous atheism. Yet today it is possible to think this even while being a well educated person, and perhaps the great majority of our intellectuals share this opinion, regarding it as a simply an incontestable fact. Nevertheless, the world is round! And the orthodox God is virtually unknown except as an object of assault in the great body of modern literature and philosophy, so that far from being alone Alfred Kazin is truly representative of modernity in his religious heterodoxy. But does a far different situation prevail in postmodernity? And is this a crucial point at which postmodernity can be genuinely recognized?

Tom

Letter 54

To Friends (October 26, 2011)

Friends,

Having just attended the fiftieth anniversary celebration of SPEP,[1] one largely centered upon Heidegger and Heidegger's metaphysics, but wholly ignoring his Nazism and nihilism, I sense a necessity of renewing a theological understanding of Heidegger—one centered on the apocalyptic Heidegger. This is the Heidegger of *Ereignis*, a word that along with *Dasein* Heidegger himself knew to be untranslatable, but it is nonetheless ultimately important for the late Heidegger, as above all manifest in *Beiträge*. *Beiträge* is the fullest statement of Heidegger's theological thinking, and his refusal to publish it during his lifetime is all too significant, for it is surely his second most important book, yet he chose to hide it from his world and his community, perhaps because it is too revealing theologically of himself. Now Heidegger's theological thinking is an apocalyptic thinking, which alone makes it impenetrable to the great majority who are closed to apocalyptic thinking, and this despite the fact that both Hegel and Nietzsche are apocalyptic thinkers, and those thinkers who had the deepest impact upon Heidegger.

We know how important Paul became for Heidegger, and already in 1920 this is the apocalyptic Paul, a Paul not discovered by New Testament scholarship until Schweitzer's great book on Paul published in 1927. The Marburg seminar that Heidegger and Bultmann taught on Paul came to have a great impact upon theology, as concretely occurring in Bultmann's *Theology of the New Testament*, which had an immense effect upon theology. Indeed, Heidegger first became embodied in America in theology and not philosophy, American theologians were immersed in Heidegger well before American philosophers, and this is manifest in a great body of American theology. Of course, the philosophical world ignores this, just as it largely ignores the theology of our great philosophers, and all of our philosophical thinkers were also theological thinkers until late modernity, and here, too, Heidegger is unique among his philosophical peers, which could be yet another reason for philosophical opposition to Heidegger.

Heidegger alone has had an immense impact upon both Protestant and Catholic theology, and while we now know that the young Heidegger was converted to Protestantism, and was enormously affected by Luther and Kierkegaard, he also was engaged in a deep quest to recover a primeval or original Christianity, one that was wholly lost in the Church and subsequent Christianity. Here, too, Kierkegaard was a great influence upon Heidegger, a Kierkegaard who has thought the passion of faith more deeply than anyone else, and who was probably the most anti-ecclesiastical thinker who ever lived. While many can understanding nothing but an ecclesiastical theology, Heidegger is the very opposite of this, which is one reason why so many judge him to be a nontheological thinker, but he is also seemingly non-theological in his refusal to think about God. But such a refusal can be found throughout not only our deeper Christian mystical thinking, but also in our deeper late modern Christian imaginative expressions, which have been judged to be atheistic by ecclesiastical authorities, just as Heidegger has been so judged by so many.

Nietzsche probably most deeply affected Heidegger in his understanding of nihilism, a nihilism that is universal as a consequence of the death of God, hence it is inescapable for us, and we inevitably deeply contend with it even if we have no conscious awareness of it. A contention with nihilism is perhaps most manifest in our late modern poetry, a poetry that profoundly affected Heidegger, making possible for him a new and radical philosophical language, and Heidegger is the only philosopher since Plato and Aristotle who along with Nietzsche truly created a philosophical language, and the only philosopher apart from Nietzsche who created a poetic philosophical language. This language is perhaps our most powerful philosophical language, but this power is inseparable in our world from a deeply nihilistic condition, a condition that even the papacy deeply contends with, and if only here philosophy and theology are inseparable allies. Of course, there are philosophers or professors of philosophy who have no understanding of nihilism, but if only thereby they must stand wholly apart from Heidegger, a Heidegger who is perhaps most truly distinctive in his ultimate conflict with nihilism.

I enclose the concluding chapter of my book, *Godhead and the Nothing*, which opens with a discussion of Heidegger and Nihilism.[2]

Yours,
Tom

Letter 55

To Friends (November 14, 2011)

Friends,

We have failed to understand Occupy Wall Street[1] because we have been attempting to apprehend it as a political, social, or economic movement, whereas it is only fragmentarily any of these things. What is most immediately important here is understanding the context of this movement, and understanding that this is a radically new context, and will fit into none of our given categories. And a fundamental ground of this virtually absolute newness is the new shallowness and alien state of our political and religious institutions, which itself is the consequence of what we theologically know as the death of God, a death that has ushered in an all pervasive absolute groundlessness, and one that is escapable only by stepping outside all of our given modes of life. Now this is the overwhelming vacuum in which Occupy Wall Street (OWS) has occurred, and as a vacuum it has made possible a new life, a new life releasing genuine energy, but in a context in which there is no possible pragmatic outlet for such energy.

At this point it is instructive to relate the OWS movement to its counterparts in the Sixties, these did have pragmatic outlets, radical political parties and organizations, religious bodies of many kinds, and genuine social movements such as feminism. This made possible a social and political and religious impact that is impossible today. Even the mass media could then participate in the movement, whereas they are only observers today, and then these movements did have an impact upon history both in America and in Europe that is inconceivable today. The very advent of a newly anonymous world is illuminating here, for this is a world in which no truly human or individual or interior distinctions are possible, a world deeply unveiled by a uniquely twentieth-century philosophy and psychology, as clearly signified by Heidegger's abandonment of that *Dasein* upon which his earlier philosophy had been centered, and by Freud's assault upon the superego and exaltation of the id.

This new anonymity is profoundly embodied in late modern literature and art, as a radically new subject is truly born, but a deeply anonymous subject, as is so decisively manifest in the ultimately subjective creations of a Picasso or a Proust. Proust and Picasso even rival Shakespeare in their creation of a comprehensive world of character or characters, but unlike in Shakespeare these characters are never actually individuals, an individuality that is impossible in late modernity, as epically enacted in *Ulysses* and *Finnegans Wake*. The *Wake* is never to be forgotten in approaching OWS, for there is an epic power and enactment in the *Wake* that is truly universal, one effecting an absolute reversal, and an absolute reversal of every established structure and power. That is a reversal that OWS seemingly seeks, but it is powerless to communicate such a reversal, or even actually to evoke it, so perhaps it is best to understand OWS as a symptom of an ultimate contemporary crisis, a symptom with no possibility of cure, but with a real possibility of itself being an ultimate summons, even if that summons is now beyond articulation.

This is not to say that we cannot now seek to articulate that summons, one that has indeed occurred in the most powerful expressions of late modernity, and if we have been distracted from these by a series of contemporary summons, it is not too late to attend to a deeper call, and even to what was once known as the call of God. It is all too significant that this call can no longer be articulated as such, hence it is absent in all of the more powerful expressions of late modernity, but its counterparts can be understood to be here present, even if they can be articulated or understood only by way of the most radical decipherment. This is a decipherment that must now be practiced upon OWS, which is perhaps easier because this is such a silent body, an enormous energy, yes, but a deep silence nonetheless, and possibly a silence inseparable from its very energy. There is no greater challenge than understanding silence, and most of all understanding a truly contemporary silence, one occurring here and here alone, and a silence that can actually be understood only through itself or its own reversal.

The silence of OWS is odd indeed, for how could such a vibrantly energetic body be so silent, is its an energy that is beyond all but the most immediate communication, and hence one that cannot possibly be actually articulated? And is this a strength or weakness of this movement, perhaps a mortal weakness, one necessitating its disappearance, and perhaps rapid disappearance? Apparently there are no true precedents for such a phenomenon, unless they have simply disappeared from history, which is surely a possibility here, and particularly so in the absence of all pragmatic possibilities. Now if OWS is ultimately made possible by the death of God, and by the unique vacuum created by that event, could OWS itself be an expression of the death of God? Here, Nietzsche is all important, that thinker who most purely understood the death of God, and who could call forth an absolute apocalypse from that death, an apocalypse that is the absolute Yes-saying of Eternal Recurrence, a Yes-saying that may well be the deepest but most hidden ground of OWS. Again and again Nietzsche has been employed as a deep key to late modernity, so it would not be odd to employ him as a key to OWS, and here we can understand an absolute No-saying that is at bottom an absolute Yes-saying. No one ever realized a deeper and more comprehensive negation of our history and consciousness than did Nietzsche, yet that negation made possible the absolute Yes-saying of Eternal Recurrence, the first total articulation of Yes-saying since the New Testament itself.

Nietzsche's relationship to the New Testament is truly extraordinary, for he is at once perhaps the deepest enemy of the New Testament and yet one of its deepest exegetes, no book surpasses *The Antichrist* in its interpretation of the New Testament, and here we have our only truly philosophical understanding of Jesus. Perhaps this book should be the grounding center of OWS, a book that may well be our only realistic understanding of compassion, and yet a compassion only made possible by an absolute assault. It is just here that Marxism is so deeply in need of Nietzsche, and apart from which Marxism has simply collapsed, for it is not in the economic arena that the ultimate weakness of Marxism lies, but rather in the uniquely human arena itself. This is the arena in which OWS is occurring, and even if this is a silent occurrence, it is a silence inseparable from noise, a noise that itself is an expression of energy, but an energy here inexplicable to itself. This is not odd or unusual, for a deeper energy is commonly inexplicable to itself, but previously it soon attracted primal spokesmen, which is perhaps impossible here, and impossible because of its absolute newness, if such a newness is actually here.

This is the primal question that OWS evokes, and this is a question inevitably calling forth illusion, indeed, is the very idea of absolute newness necessarily inseparable from illusion, and even from total illusion? Total illusion may well be rare, but Nietzsche could know it as occurring in the very advent of Christianity, which he could know as the most catastrophic event in history. For Nietzsche knew Christianity as an absolute reversal of a noble morality, or of all higher humanity hitherto, and that reversal is the very essence of Christianity, even if this reversal is a reversal of Jesus! Nietzsche knew Jesus as the very opposite of Christianity itself, for Jesus is that Yes or affirmation who is the opposite of an absolute No, a No that is nowhere so purely embodied as it is in Christianity. Nietzsche is perhaps the most purely dialectical of all thinkers, or of all Western thinkers, and consistently dialectical as even Hegel is not, so that the No that Nietzsche knows is inseparable from Yes, and the absolute Yes of Eternal Recurrence is inseparable from the absolute No of eternal return. If it was his own thinking that drove Nietzsche insane, and even his deepest thinking, it is Nietzsche who is our deepest pathologist, the thinker who most deeply understands an absolute No.

This, too, must never be forgotten as we approach OWS, for OWS does attempt to evoke an absolute No, and shallow as it may well be, it is perhaps most powerful in saying No, and saying No to all established power. This is a No created or first evolved by the prophetic revolution of Israel, one ignored by all philosophy except for Nietzsche, and if only at this point Nietzsche is of ultimate importance, and certainly of importance in understanding OWS. Now it is not insignificant that OWS has had such a universal impact, perhaps the first time that a prophetic No has been heard throughout the world so rapidly, and if this is a genuinely prophetic No, it is inseparable from a profound offense. If this offense is not a strong one that would be a decisive sign of the weakness of OWS, but on the other hand if it is strong, that would be evidence of the strength of OWS, although it may well be that the absence of pragmatic outlets makes it impossible to resolve this question.

Tom

Letter 56

To Friends (December 26, 2011)

Friends,

I have come to think or suspect that we commonly lack agreement as to just what apocalypse or the apocalyptic is, leading to a fundamental block in communication. So this is an attempt to call this forth, and I begin with what is generally accepted as biblical apocalyptic texts. Historically, these begin with the prophetic books, and perhaps most so the Book of Isaiah, including first, second, and third Isaiah. Then there are a number of noncanonical apocryphal works, and above all so the Dead Sea Scrolls, reflecting the advent of a genuinely apocalyptic movement arising early in the second century BCE. Christianity arises out of this movement, and with the exception of the Fourth Gospel and many of the epistles, the New Testament is an apocalyptic book, and above all so Paul and the Synoptic Gospels. This was not historically discovered until the end of the nineteenth century and many including myself think that this is the gravest theological challenge of late modernity. In this perspective, Jesus is a fully if not a purely apocalyptic figure, and if only thereby virtually all of our theology is alienated from Jesus.

Then there have been truly major apocalyptic movements in Western history, such as the medieval Joachites who are reborn in later history, and then most of the bodies in the Radical Reformation. Here, Blake and Hegel are indirect descendents of the Radical Reformation, and it is important to accept Blake and Hegel as being deeply apocalyptic, even if they are vastly distant from what is commonly understood as apocalypticism. Indeed, this is a major semantic problem for us, since there is a vast distance between common or popular apocalypticism and deep apocalypticism, as most openly manifest in William Blake. Obviously Blake and Hegel cannot be confused with popular apocalypticism, as perhaps most manifest in their profound enactments of the death of God, but thereby they reenact the Crucifixion, and profoundly renew the apocalyptic Jesus, who had been lost in the great body of Christian tradition.

It is fascinating that we have been given so little apocalyptic theology, except for that produced by apocalyptic sects, so that apocalyptic theology is virtually unknown in the theological world. Blake scholarship has certainly produced rich and exciting interpretations of Blake's apocalyptic vision, which is purely heterodox from an orthodox perspective, but so, too, is Hegel purely heterodox, and heterodox as an apocalyptic thinker. Now if German idealism itself can be understood as being apocalyptic, then it is thereby a profound challenge to theology, and is so insofar as it is genuinely theological. But in being apocalyptic it is a renewal of the New Testament, and also a renewal of early or primitive Christianity, and as such an ultimate challenge to all established or manifest theology.

Thus genuine or deep apocalypticism is the greatest challenge to theology, and also the greatest challenge to our common or given understanding. But it is also important to understand that apocalypticism has played a fundamental role in all of our world revolutions, beginning with the English Revolution, and then exploding in the French, Russian, and even Chinese Revolutions. Hence apocalypticism can be understood as the most powerful purely negative power in world history, and it has transformed the world more than any other power, for even the scientific revolution of the seventeenth century can be understood as an apocalyptic event insofar as it brings an end to an old world and ushers in an absolutely new world. Never must it be forgotten that a genuine apocalyptic ending is simultaneously an absolute beginning, this already occurs in the advent of Christianity. And here an opposition to the God of Law or Temple or Kingship is an opposition to the God of the old creation or the old aeon, and a witness to the God of the new creation or new aeon, a new aeon enacted in Jesus' proclamation and enactment of the Kingdom of God, which is just why that "kingdom" cannot be understood as either kingdom or reign.

In this perspective, Jesus is the most revolutionary figure in world history, and Christianity the most absolute challenge to all established power and authority, but if Christianity was absolutely transformed in becoming an established and ruling Christianity, that transformation is the greatest in world history, and one demanding a revolutionary transformation of all given or established Christianity. There is even an awareness of this in the Second Vatican Council, which is a virtual reversal of the First Vatican Council, but that very reversal can be understood as a renewal of primitive Christianity, a primitive Christianity that was a truly explosive force, and by necessity so given its deeply apocalyptic ground. Accordingly, genuine Christianity is an explosive Christianity, and its conservative expressions a decisive sign of an ultimate counterrevolution, and even the greatest counterrevolution in history. Perhaps it was Nietzsche who understood this most profoundly, and just as the Jesus of *The Antichrist* is the absolute opposite of Christianity, this is a Christianity that has absolutely inverted or reversed Jesus, a reversal that is the greatest reversal in world history. So it is that Nietzsche could embody Jesus only by realizing the actual death of the uniquely Christian God, the only God capable of such a death, and one whose death releases an absolute nihilism.

Yet this is a nihilism not only inseparable from apocalypticism, but one actually realized by apocalypticism, so that it is impossible apart from apocalypse itself. Thus the death of God is an apocalyptic event, and an ultimate apocalyptic event, and therefore it cannot be erased or annulled or dissolved. Now this is just what genuine apocalypse is, so that it is inevitably revolutionary, consuming everything in its wake, and in its expression in the Christian epic tradition, as occurring in Dante, Milton, Blake, and Joyce, it envisions and enacts an absolute world-transformation, and a world-transformation of totality itself. Genuine apocalypse is inevitably and necessarily an absolutely new totality, but it nonetheless is an actual totality, and a totality that is all in all. There can be no standing outside of this totality, or no standing that is not a falling into abyss, and a falling into an ultimate or absolute abyss. Thus apocalypse poses its own Either/Or, either new aeon or old aeon, either Spirit or Flesh, either Heaven or Hell, but this absolute antinomy finally realizes a *coincidentia oppositorum*, or what Blake comprehensively enacts as a marriage of Heaven and Hell.

Tom

Letter 57

To Friends (February 11, 2012)

The Unique Ground of America[1]

The historian Perry Miller[2] understood that America was the first founded nation, the first nation to be born as the consequence of an actual decision or will, as opposed to being born as the consummation of a social or historical process. America reveres those founding fathers who enacted this decision, and this founding was the first such founding in history, thereby giving America a truly unique identity. Europeans can know America as a nation without a history, and it does lack everything that most peoples know as tradition, here there does exist a uniquely American void, and a void ever threatening an ultimate disruption of America. Such a disruption occurred in the American Civil War, the most violent civil war in history, and a civil war only made possible by violent fissures in American life, and not only class divisions that were less than those in Europe, but a deep polarization throughout society, as most manifest in the slavery controversy. No doubt language divisions were then important, too, and perhaps what is most uniquely American is the American language itself, and as H. L. Mencken's great book[3] on the American language demonstrates, what is truly distinctive of this language is its dissolution of all given rank and station, or all given linguistic hierarchy, thus for the first time creating a language world without any hierarchy whatsoever, and this has been an offense to English speakers elsewhere.

There is only common speech and common language in America, hence America has been without a genuine aristocracy, and a distinctly American fiction has given itself to a dissolution of the illusion of an American aristocracy, as so profoundly occurring in Faulkner[4] and Henry James.[5] Is America the first society to be without an aristocracy, a society so banal that it is money alone that determines rank, and if it is money that is the true "God" of America, is that too unique? William Blake, our most revolutionary modern prophet, in his first epic, *America*, first enacted the death of God, an enactment occurring in what Blake knew as a revolutionary America, an America confronting and slaying the God here named as Urizen, who is a uniquely modern absolutely alien God. So, too, in America's greatest epic, *Moby Dick*, the White Whale is an embodiment of the absolutely alien God, a God so alien and empty that it is here enacted as the Nothing, an initial enactment of the death of God that soon becomes overwhelming in the late modern imagination. Thus America is a pioneer in ways seemingly unknown to itself, as the most religious of modern nations can be understood as having inaugurated atheism, or inaugurated a deeply modern and profound atheism that has no historical parallel.

Yet if America is the first founded nation, atheism can be understood as being essential to that founding, or if not atheism, a dissolution of the deepest ground of tradition itself, a ground that is commonly named as God. The American founding fathers were sons of the Enlightenment, an Enlightenment that is nothing less than the inauguration of modern atheism, as so purely embodied in Hume, who can be known as the greatest Enlightenment philosopher. But the Enlightenment both in Europe and America impelled a profound religious reaction, one fueled in America by a uniquely American pietism, a pietism without a genuine theological ground, and one given to passionate emotional convulsions. These are deeply understood by America's greatest philosopher of religion, William James,[6] a psychologist and philosopher who himself was a victim of a deeply interior pathology, a pathology deepening his own thinking. This opened James to what is perhaps a uniquely American void, one enacted again and again in American drama and literature, and most clearly so in the dramas of Eugene O'Neill,[7] who is commonly known as America's greatest playwright.

O'Neill's deeper dramas can be apprehended as being profoundly nihilistic, perhaps purer enactments of the Nothing than any other drama apart from *King Lear* and *Oedipus at Colonus*, and if this is a deeply American Nothing, it is so in its very actuality as the Nothing. While pragmatism is commonly associated with America, the pragmatism of William James is not a common pragmatism, and is not because it is open to depth itself, as manifest in *The Varieties of Religious Experience*. Here, it is only a once-born or healthy-minded religion that James cannot enter, which is perhaps essential to his so deeply entering the sick soul, and just as James himself is a sick soul, he thereby becomes open to a deeper America. This is the America that is enacted in the most powerful plays of O'Neill, but it is a nihilistic America, an America stripped of all ultimate grounds, as here both action and speech are ultimately empty, as these dramas continually enact a nihilistic destiny, and one calling forth a uniquely American destiny. No doubt that emptiness is manifest in American Capitalism and American Imperialism, and if these are truly pragmatic expressions of America, they are nonetheless genuinely empty expressions, and perhaps all the more deeply American because of that.

Thus we are challenged to understand a uniquely American emptiness, one certainly embodied in the American imagination, but in American religion, too, a uniquely American religion, but one not yet understood as such. Now just as America is the first founded nation, its religion might be understood as the first founded religion, or the first to arise from a genuine void, a religious void deriving from the arrival of a truly secular world, one demanding a separation of religion not only from the state, but from the more powerful expressions of culture. The first separation of church and state in America must be understood in this context, and this is an enormous transformation, as not since primitive Christianity had there been a church truly independent of the state, except for the most radical sectarian expressions of Christianity. It is not insignificant that these have flourished more in America than anywhere else, and that Christianity is far more diverse in America than elsewhere, just as a nonecclesiastical Christianity is perhaps most powerful or most common in America.

Consequently, one must inquire if there truly is a fully secular America, or an America that actually is empty of religion, or genuinely atheistic in the common sense. It is remarkable that there are virtually no major expressions of the American imagination that are atheistic in this sense, and while this is true of all of the primal expressions of the imagination, the American imagination has from its beginning been free of an established religion or an established orthodoxy, and thus been heterodox as has been no other imaginative tradition, even if such heterodoxy has led to a deep isolation of the most powerful expressions of the American imagination. This is perhaps most openly manifest in the common expressions of American religion, these are virtually bereft of a genuine imagination, except perhaps for a peculiarly American simplicity, one reflecting a uniquely American innocence, an innocence free of the actualities of history and experience. But this, too, is a reflection of a profound emptiness in America, and a uniquely American emptiness, and one that can even be understood as being the initial expression of a post-Western society and civilization.

This emptiness can be apprehended as a uniquely American nihilism, one extraordinarily powerful in its pragmatic effect, but one impossible apart from the ultimacy of nihilism itself, or apart from a dissolution of all ultimate grounds. Now if America created a uniquely modern corporation, one grounded in financial power alone, and wholly independent of all tradition, this is a financial power that never previously existed. This is why the greatest philosopher of capitalism, Karl Marx, could not understand American capitalism, and perhaps could not understand it because Marx is innocent of a nihilistic sensibility, and thus closed to a nihilistic capitalism. That is the deepest distance between Marx and his twentieth-century successors, successors who beginning with Lenin were profoundly nihilistic, thus making possible an absolutely nihilistic Stalinism, and a Stalinism whose most ultimate enemy was not Nazism, but rather America itself.

The conflict between America and the Soviet Union can be known as a conflict between two ultimately nihilistic powers, and just as both Russian and American literature deeply enact a uniquely modern nihilism, each evolved truly empty societies, or societies empty of all genuine history and tradition. Nothing like this conflict occurred in previous worlds, a conflict that is Manichean as an absolute conflict between truly opposite powers, and truly Manichean also in the parallel identity of these opposite powers. Peoples independent of these powers could know each as being either empty or nihilistic, and just as a hatred of both Russia and America evolved that was truly new, such hatred is a genuine response both to a new America and a new Russia. Has that Russia now disappeared, a disappearance accompanied by a disappearance of that America, or far rather an even deeper epiphany of that America? This is an ultimate question facing us today, but is it a question whose response can occur only in a far emptier America, the emptiest of all Americas, and yet one wholly unaware of that emptiness?

American innocence is truly remarkable, it can occur even in an advanced and sophisticated society, but it is most pervasive in American technology, the first technology to become a total technology, and thereby to create a truly new technological world. Even farming in America has become totally technological, and if this effected the second agricultural revolution, there are parallel transformations in society itself, and in the advent of a technological imagination, as expressed in music, painting, and architecture. Thereby is born a truly new nihilism, a fully comprehensive nihilism, one pervading all worlds, and even creating a new interior world, one foreseen by a Dostoevsky or a Nietzsche, but only actually initially enacted by a wholly technological America. Hence what an overwhelming difference between twentieth-century and eighteenth-century America, a difference seemingly impossible to bridge, but obviously there was an historical transition between one and the other, although it is extraordinarily difficult to understand that transition. A genuinely secular world was perhaps born in America, but the American secular world is never dissociated from a religious world, which is what is most unique about the American twentieth century, yet that century is nonetheless a nihilistic century.

An integral clue to this America is jazz, the most American of all American arts, and also the most universal, for it has embodied America throughout the world, and become a source of ultimate joy. Initially jazz was responded to with a real opposition, as though it were a nihilistic assault upon the world, and this occurred at a time when American anarchism was seemingly powerful, so that jazz could be understood as an expression of this anarchism. While this appears to be fantastic today, it once was seemingly common sense, and it is a witness to the power of jazz, a jazz not only born in America, but clearly the expression of an American energy and life, a life threatening to consume everything about it. This, too, can be understood as an expression of an American nihilism, a nihilism consuming all culture and tradition, and emptying everything in its wake, even if such emptying can release a truly new energy, such as the unique joy created by jazz. Indeed, jazz is a decisive witness to this emptying, one that is apparently irresistible, and just thereby universal, so that American nihilism can be known throughout the world, and recognized in those distinctively American creations, mass advertising, rock music, and movies for the masses.

It is foolish to think that nihilism can only be known or experienced by a few, it is rather the advent of the masses that makes possible a uniquely modern nihilism, many know that advent as occurring in the French Revolution, and being renewed in the Russian Revolution, but it is truly universalized by America, and if not by the American Revolution, then by America itself. Now it is remarkable how little our nihilism is discussed, for while often mentioned it is seldom actually meditated upon, and this despite its prominence in Nietzsche and Heidegger, and its embodiment in our greatest late modern literature and drama. Most remarkably there is little confrontation with nihilism in our theology, and this despite a common theological recognition of nihilism as a consequence of a uniquely modern enactment of the death of God, a nihilism often employed in theological assaults upon atheism, but never employed with a genuine inquiry into nihilism. Is nihilism itself an absolutely forbidden subject, one closed to all genuine inquiry, and most so in an age that many can recognize as a nihilistic age? There are powerful assaults upon Hegel that recognize Hegel as the father if not the creator of nihilism, thereby recognizing a uniquely modern philosophical tradition as a nihilistic tradition, as above all occurring in its understanding of God. This is a truly ironic thesis, for there is no philosopher for whom God is more absolute than Hegel, nor a philosophy more centered upon God than Hegelianism, yet it is Hegel who first philosophically enacts the death of God, and many can understand this as the very inauguration of a fully modern philosophy. This is the death of God that can be understood as the inauguration of nihilism, but Hegelian philosophy is seemingly the very opposite of nihilism, and is so as an absolute idealism, and an idealism in which an absolute reason is all in all. But is that the very idealism that is an ultimate nihilism, and even a nihilism in its virtual deification of Reason or *Vernunft*, a reason replacing God, and even replacing God in the name of God, or in the name of the Crucified God?

Just as Hegel made Marx possible, and ironically made Kierkegaard possible too, Hegel is an ultimate founder, but is he the founder of nihilism? To accept this is finally to affirm that a uniquely modern world is a nihilistic world, and this has long been a conservative thesis, one soon following upon the French Revolution itself. No one understood the French Revolution more deeply than Hegel, and he could know it in the *Phenomenology of Spirit* as the historical inauguration of the death of God, an inauguration that is the historical destiny of the West. Is that a destiny that is a nihilistic destiny, and nihilistic as an enactment of the death of God, an enactment finally ending every ultimate or absolute ground? For Hegel, the death of God inaugurates a total apocalypse, so that the modern realization of the death of God will culminate in the New Jerusalem, a Kingdom of God totally realized here and now. Hegel is our most profoundly apocalyptic thinker, and far more apocalyptic than Augustine or Luther, and it is just as such that he is our most revolutionary thinker, but in being that revolutionary thinker he can make possible an understanding of his own philosophy as the creator of nihilism, so that it is even possible that this philosophy is our most Satanic thinking.

Hegel could know America as the land of the future, and Hegel did once dominate American philosophy, just as he had an enormous impact upon American drama, for Hegel is the supreme philosopher of tragedy, even as he could understand comedy in the *Phenomenology of Spirit* as being profoundly Christian, and understand Greek comedy as culminating in Christianity. Is it possible to understand nihilism in a comic context, and in an Hegelian comic context, so that nihilism is finally an absolute inversion, reversing everything whatsoever? This would be a reversal that is finally an absolute transfiguration, but a transfiguration inseparable from an absolute negation of the totality of the given, a given that Hegel was the first philosopher to know, a knowledge only made possible by its absolute negation. Hegel is the supreme philosopher of negation, creating absolute negation itself, which he could know as occurring in and as the Crucified God, an occurrence that *is* absolute apocalypse. Indeed, Hegel is our only philosopher of the Crucified God, unless this is true of Nietzsche, too, and above all of that Nietzsche who is the greatest master of nihilism.

Nietzsche could know theology as a pure nihilism, and know the uniquely Christian God as the will to nothingness pronounced holy, and if Nietzsche appears to be infinitely removed from Hegel, he nonetheless is along with Hegel the primal nineteenth century influence upon twentieth-century philosophy. Derrida and Deconstruction truly integrate Hegel and Nietzsche, which at least demonstrates that this is possible, and this does bring a new identity to our nihilism, which can then be known as a universal nihilism. A universal nihilism appears to be a contradiction in terms, for how could nihilism be universal except in a wholly insane world, and while many know our world as being such insanity, is it actually possible to be conscious of that insanity? There are investigations of suicide that suggest such a possibility, and if Camus could make an immense impact by centering upon suicide, it may well be suicide that most illuminates our nihilism, or most draws forth its concrete identity for us. Dostoevsky, too, has a profound understanding of suicide, and his Kirilov[8] may well be not only the patron saint of suicide, but the patron saint of nihilism itself.

It may appear to be odd to associate nihilism with a patron saint, but it is not odd if Nietzsche is that saint, and while few can know Nietzsche as a saint, he can be known as such in the context of his absolute courage, a courage to plunge himself into the depths of nihilism, which alone could make possible his profound understanding of nihilism. All too significantly that understanding is confined to his notebooks, and just as Nietzsche's notebooks can be known as our greatest notebooks, or our greatest along with Leonardo's and Kafka's, the understanding of nihilism here is profoundly offensive, and offensive because it is so actual and real. There may be positivists among us who scorn such understanding, but this is an understanding that ends all possible positivism, and ends it by understanding it as yet another expression of nihilism, a nihilism that truly is universal. Perhaps silence alone is the only proper response to nihilism, and while such silence may be extraordinarily difficult to achieve, this is not only possible but actual if grace is everywhere, and perhaps nothing is a greater summons to grace than is nihilism itself.

Thomas J. J. Altizer

Letter 58

To Friends (March 2012)

Friends,

William Hamilton[1] died on Leap Day of 2012, peacefully in the company of his full family, after having had a truly unique effect upon American theology. It was Bill who in the Sixties created the scandal of the advent of a death of God theology, and did so in a CBS national television production of a series devoted to the death of God, featuring films, theatre, and forums, and Bill's own guiding commentary on the death of God.[2] This is certainly the most powerful of all theological television programs, and in its perspective the Billy Grahams of our world are very weak indeed. Bill Hamilton had an extraordinary talent for public communication, and only Bill could speak genuinely theologically to a broad public, and the real center of Bill's theology was Jesus and the praxis of Jesus, which for the first time were communicated to a vast audience. (This might have occurred in Rome on RAI television, the BBC of Italy, but the program on the death of God in which I participated was commissioned when there was no Italian government, and when one returned, the program was canceled.) It is very difficult today for most people to realize the publicity given to radical theology in the Sixties, even the *New Yorker* could devote two whole issues to it,[3] and there was simply no magazine or newspaper that did not feature it at one point or another, and in Easter of 1966 it was the cover issue of *Time*, which for the first time published a cover wholly in black, asking "Is God Dead?" and to this day this is the best selling issue of *Time*.

Bill was a great preacher, and he could preach in our common language, there was nothing academic about it, even if he was an excellent scholar. While Bill had a genuine loyalty to a liberal or radical Baptist tradition, and was a descendent of the Radical Reformation, he also had a deep ground in literature from Shakespeare through Melville, and made substantial contributions to literature and theology. Once literature and theology was a very promising field, and if it is largely moribund today that is because of a contraction of theology that has occurred since the Sixties, one inseparable from a new dominance of theological conservatism. But are the Sixties really that distant from us? Could Bill Hamilton's death remind us of his way in such a manner as to make possible a renewal of radical theology, and a renewal of radical theology in our world? Is it possible that genuinely evoking a radical theology in a common language, and a language that has an immediate or passionate impact, is a decisive way to that renewal?

This is how we can best honor Bill today: not simply to remember him but to renew him, a renewal that his life made both possible and actual, for even if his life was a unique one, it was also a common one in the best sense of that word, as Bill was a theological everyman who could be a model for many. Arrogance was wholly alien to him, just as was any esoteric understanding of theology, no one was more committed than Bill to a theology for everyone, and a critical theology for everyone, something that we have not had since the Puritan era. While it may appear odd to associate Bill with Puritanism, this is not odd if there is a living American Puritan tradition, and one diverse enough to include a Hamilton, and if no theologian has been more American than Bill, this is an Americanism inseparable from Puritanism. But inseparable from a genuine Puritanism, and no chasm is greater in our world than that between a common understanding of Puritanism and a critical understanding of Puritanism, for genuine Puritanism is as powerful as any other force in our history.

Recall that the Puritan ideal is one of making the Christian Word actually and fully incarnate in our world as a whole, while Bill initially absorbed this idea at the Union Theological Seminary from Reinhold Niebuhr, he gradually made it far more concrete than it has been at our seminaries, and this entails an ultimate loyalty to Jesus. But most paradoxically it simultaneously entails an ultimate acceptance and affirmation of the death of God, here Bonhoeffer had a profound effect upon Hamilton, and perhaps a deeper effect upon Hamilton than upon any other theologian. Such a dissociation of Jesus and God appears to be very odd indeed, but it was already primal for William Blake, just as it is in a radical American tradition, a tradition for which William Hamilton has been our primary contemporary spokesman. Since the Sixties a fundamentalism of innumerable varieties has become so pervasive as to wholly obscure such a tradition, but Hamilton deeply embodied it, and not only embodied but actually spoke it, and spoke it with compelling power.

Bill was not only a great preacher but a great teacher, too, and nothing is a more powerful witness to him than his students, who encompass an extraordinary theological breadth, and one going well beyond theology. While I visited him at each of the institutions where he taught, I sensed that he could never find a genuine academic home, and not even when he was a dean at a major university. This may well be true of a great many today, but Bill embodied one of our most unique identities, and perhaps above all so because he spoke a language to which almost anyone could respond. Therein he was teacher and preacher at once and altogether, as even his writing is homiletic as little writing is, and far more homiletic than our open or literal homiletic writing. Since we often spoke in public together I had an immediate experience of this, and he had a command of the media as did no other theologian; often he counseled television directors, and this even occurred with the highly professional directors in Rome.

It has been a long time since preaching was taken seriously in our world; indeed, few now have an awareness of the power of preaching, or a realization that Christianity originally had an overwhelming impact because of the radically new power of its kerygma or preaching. Bill was dedicated to renewing that kerygma, but renewing it through a demythologizing that he realized independently of the Bultmannians, for Bill was not deeply affected by New Testament scholarship, even if he wrote New Testament commentaries. His was an American and perhaps uniquely American way of renewing the gospels, a deeply oral way, in which one resays the gospels in a fully contemporary language, and the gospels speak themselves in that language. Oddly enough this was not unrelated to Bill's profound love of Shakespeare, and if he could reenact Shakespearean drama in his own language, this, too, was homiletic, as a strange conjunction of Shakespearean language and the language of the gospels now occurs. Again, this appears to be very odd, but it was incredibly powerful, and powerful in an immediate way.

Tom

Letter 59

To Friends (March 7, 2012)

Friends,

Yesterday while I was talking about theology to a *New York Times* reporter I remarked that just as the middle or moderate is dropping out of our politics, it is dropping out of our religion, too, as witness declines in liberal Protestantism, liberal Catholicism, and Reform Judaism, and possibly of liberal Islam and Hinduism also. This very deeply affects our theological situation, one that is now dominated by the Right, but it simultaneously opens new paths for the Left or for radical theology. I sense that there are now more radical theologians than ever before and that this is reflected in our academic meetings, and also to a significant extent in our publications, which are possibly more radical than ever. Maybe I can induce *The Christian Century* or someone to investigate this, or perhaps we should form a radical theology society that could meet simultaneously with the annual meetings of the SBL and the AAR.[1] Or could there even be a radical theology section or group in the AAR? Somehow straights always seem to gain control of these groups although Ray Hart could certainly guide us in this arena.

There is an interesting essay on God and Caesar in America in the current issue of *Foreign Affairs*[2] emphasizing the deep impact of the Sixties, and the subsequent expansion of evangelical Christianity and the virtual collapse of liberal Protestantism and "churchgoing" Catholicism. The rise of the religious right that now occurs is here identified as a common theme in US history, but today it is occurring with a claim that there is a God gap between Republicans and Democrats, and fewer and fewer Americans now identify as secular Republicans or religious Democrats. And now politics appears to determine religious practice, as formerly religious Democrats (except for African Americans) have drifted away from church, and formerly unobservant Republicans have found religion. Surveys have found that the strongest predictor of a Republican becoming a Tea Party supporter is whether he or she evinced a desire to see religion play a prominent role in politics.

Yet there has been a great growth in the number of people whom sociologists call "none," those who report no religious affiliation. Historically, this made up a constant 5–7 percent of the American population, and even did so in the Sixties, but in the early nineties, nones made up 12 percent of the population, and by 2011 they were 19 percent. This is a huge demographic shift, and now younger people are far more likely than their parents to reject all religious affiliation, as 33 percent of them do. The best evidence indicates that this dramatic generational shift is primarily in reaction to the religious right. Democrats and progressives are much more likely to become Nones than Republicans, and the religious defections were concentrated specifically among those Americans who reported the greatest discomfort with religion infused politics, regardless of their own partisan loyalties. All too ironically an evangelical revival has led to more religious defections than ever before!

Few intellectuals have ever encountered an evangelical, but as one who once debated them I can give witness to their passion, and a passion directed to the King James Bible, although it is seldom accompanied by any actual knowledge of the Bible. As I used to joke with my Emory students, how can you believe in the Bible if you never read it, but as one who taught introductory undergraduate courses in the Bible for over twenty years, I can testify to a common student excitement with the Bible once they actually enter it, and it is always strange country for all of them. This is an important point that is seldom acknowledged, and even if I inspire fury when I publicly declare that the death of God theology is more Biblical than orthodox theology, I am unaware of ever having been defeated in argument on this point, although it has occasioned threats of my murder. I deeply love the King James Version (KJV), and am aware of how sacred it is to evangelicals. Once on a television program in Atlanta I mentioned a fundamental error in the KJV, and their switchboard exploded as it never had before. So I adopt the strange path of confronting evangelicals with the Bible, and above all with the proclamation and praxis of Jesus, drawing their attention to the primacy here of the Kingdom of God, and a Kingdom of God that has never entered our common or established theology.

Tom

Letter 60

To Ray L. Hart (May 13, 2012)

Dear Ray,

I don't know whether or not you are aware of it but for many months I have been writing about Radical Catholicism in a quest for the essence of Catholicism, which I sense is unknown although profoundly embodied in Leahy's "thinking now occurring for the first time."[1] Several rough papers have come out of this, and I enclose the most recent one. In many ways the whole project is inspired by Erich Auerbach and his radical understanding of *mimesis* that comes out of his great book on Dante,[2] and in the enclosed paper the challenge is to bring out the polar expressions of *mimesis* in Dante and Joyce, and in such a way as truly to integrate them, and hopefully to draw forth a uniquely Catholic *quia*.[3] As far as I am aware no Catholic theologian has taken up such a project, but unlike Catholic theology I believe that the true essence of Catholicism is unknown, and thus far only realized imaginatively. The real challenge lies in understanding Catholic realism, a realism only released by an absolutely new world, hence a realism that is finally an apocalyptic realism.

Pius IX[4] is a key figure here, as is a uniquely modern papal Catholicism, one that created a new natural law condemning contraception, although this was wholly ignored in the recent controversy about contraception. But this paper centers upon the dogma of the Immaculate Conception, wholly conjoining it with the dogma of papal infallibility, and attempting to show that these dogmas were only made possible by the ontotheological event of the death of God, which shattered all order and authority, thus demanding an absolutely new authority. Here there are genuine parallels between Catholicism and modern totalitarianism, the former making the latter possible, but in such a way as absolutely to subvert it, although this can only be the subject of a future study. The great weakness of the common studies of Catholicism is their removal of Catholicism from the actualities of the world, rather than calling forth the role of Catholicism in creating these actualities.

While there are excellent studies of Dante and Joyce, I am unaware of any genuine attempt to integrate them, and if we understand Joyce as imaginatively expressing the dark underground of Catholicism, thus as it were recreating Rabelais, but recreating Rabelais in an absolutely common language, and a common language that speaks us as does no other language, this is a language that is finally coincident with the language of Dante's *Commedia*, even if such a coincidence is a *coincidentia oppositorum*. Everyone can understand Dante as the greatest poet of Catholicism, but few realize the great heterodoxy of the *Commedia*, which is just where it is coincident with the great epics of Joyce, the papacy did not condemn Dante for nothing, but rather for his equation of Church and Empire, an identity embodied in the *Commedia*, and one releasing an ultimate heterodoxy that is an absolutely new Catholicism. Genuine Protestants can understand this, and thereby understand Catholicism as a *horror religiosus*, one without parallel in the world, and an actual incarnation of evil and darkness. But that darkness is actually real, and finally inseparable from absolute light, an absolute light that is nowhere so deeply celebrated as in the *Commedia*, and an absolute light that is inseparable from Catholicism itself. What is most startling about Catholicism is that its light and darkness are both so actually real, an actuality celebrated most purely in the Eucharist, a uniquely Catholic Eucharist that is an *anamnesis* or renewal of the Breaking of the Body of Christ, which is nothing less than an *anamnesis* of the Crucifixion itself, and of that Crucifixion that *is* Resurrection.

Unfortunately almost all of our understanding of the Eucharist removes it from actuality itself, thus understanding it from within the perspective of the history of religions, and thereby wholly losing its uniquely Catholic ground. Medieval liturgical developments have been understood as a paganization of the Eucharist, as in the liturgical movement itself, hence the calls for a return to the early liturgy, returns actually enacted in the Second Vatican Council. Indeed, the relation between the First Vatican Council and the Second Vatican Council is an intriguing one, the first wholly reactionary and the second genuinely progressive, and this polarity is now writ large in the Catholic Church, and if nothing else this does create new theological possibilities. So the great Catholic question of our time is whether or not Catholicism can again genuinely renew itself, is there now any possibility of anything like the Gothic Revolution, one that not only produced an Aquinas, an Eckhart, and a Dante, but a whole new world that became the modern world? Despite the fact that modernity is the greatest enemy of Catholicism, Catholicism created modernity in the Gothic Revolution, as great an historical revolution as has ever occurred.

Tom

Letter 61

To Friends (June 19, 2012)

Friends,

Charles Stein's *Persephone Unveiled*[1] is a truly remarkable book in its full conjunction of poetry, philosophy, psychology, and theology, but perhaps it is most daunting in that it does not give offense, and this despite its Jungian perspective. Freud and Jung are paradigms of a uniquely modern offense, the one offending our deepest ground and the other offending all our sense of decorum and order, the one being overwhelmingly influential in modernity, and the other being almost wholly peripheral, or peripheral outside of sectarian bodies. Stein is a major figure in a small body of theogonic poets who give witness to a Divine totality in a wholly secular world, a world that has erased every sign of the Divine, but paradoxically thereby has made possible a total embodiment of the Divine, as perhaps most decisively manifest in our greatest modern poetry.

One of the true challenges of that poetry is its homiletic power, one that is undeniable but almost impossible to understand, first, there is the problem of homiletics itself, the least developed and most invisible theological discipline, and yet the most powerful of all theological disciplines or modes in its profound impact, and not least so in modern poetry. Clearly our literary critics are wholly baffled by this impact, even if many poets freely confess it, but they do so in a world in which theology itself is wholly circumscribed or wholly hidden, and banned from those worlds in which literary criticism occurs. Many if not most critics are deeply embarrassed that great modern poets such as Yeats and Eliot are so profoundly religious, and virtually none of them can deal with great religious poetry, as though this is a truly forbidden language, or one that can only be entered with a loss of all status and respect.

Perhaps this situation is most revealing in its treatment of epic poetry, the most religious or sacred of all of our poetic genres, but the one most ignored in our academic and critical worlds, and so much so that it is not even possible to discover a serious conception of epic in our critical languages. Does anyone even know what epic is? And yet it is perhaps the most influential of all of our poetic languages, the one that has most profoundly shaped our cultures and societies, as the very names of Homer, Virgil, Dante, Milton, and Blake make manifest. Blake's greatest epics, *Milton* and *Jerusalem*, are virtually unknown to everyone, and this without any embarrassment at all, just as there is little embarrassment that the Bible itself is unknown in our common critical and academic worlds.

Of course, this is the consequence of the advent and dominance of the most secular world in history, a dominance that ironically is even hidden from itself, as we have so little critical understanding of secularity, for even our great studies of secularism as in Weber and Durkheim are hidden from most of us. But one question does tantalizingly arise again and again, and that is whether or not there is any genuine or great secular poetry, and this is the context and perhaps the only one that will not allow us to forget Lucretius.[2] Yet whom else can we name? Is there a major and modern poet whom we can know as a genuinely secular poet? Here, Larkin[3] is often mentioned but almost mentioned alone, as though this is a genuine anomaly, or a true defiance, and could this be because a genuinely or a truly secular poetry is an oxymoron? And an oxymoron because genuine poetry is wholly alien to the secular, and is simply inseparable from a profoundly sacred ground?

One problem many have is that they deeply love or are bound to both poetry and the secular, cannot imagine themselves as being anything but secular but nonetheless love poetry, and cannot imagine a genuine love of poetry that is not a secular love. Here, Nietzsche is a primal figure, that ultimate atheist who loves poetry so deeply, and was more influential upon the twentieth century than perhaps anyone else, a world in which both poetry and atheism are all in all. So it is that our modern sacred poets have only been sacred poets by way of a profound conflict with atheism, a conflict unique to the modern era, and it is even in our poetry that our deepest expressions of atheism occur, an atheism that is not even born until Blake, and then almost immediately becomes all pervasive. Yet here lies the deepest paradox of modern poetry, it is profoundly atheistic and profoundly sacred at once, as an absolutely new poetic homiletic is created, one with enormous sacred power and a truly and fully atheistic ground, even if this is an absolutely new atheism, but one already enacted in Blake, and in Hegel and Nietzsche too.

One great gift of modern technology is oral recording, so that later generations can actually hear our major poets, and who could forget the voice of a Yeats[4] or an Eliot[5] once one has heard it, and therein heard a truly sacred voice? And this is undeniably a homiletic voice, far more homiletic than the voices we commonly hear in the pulpit, and more so if only because a new sacred language is now being spoken, and not an archaic language as are our biblical languages, but a contemporary language that speaks us in its language, and speaks that which we are incapable of saying ourselves. This is a sign of genuine poetry, that it actually speaks that which otherwise cannot be spoken, and yet it is recognized as genuine speech as it occurs, and recognized by all within its hearing.

Such speech engages us as does nothing else, an engagement that is an ultimate challenge, and one can even evaluate poetry at least for oneself in terms of the challenge that it poses. Homiletic power is only genuine insofar as it occurs within its hearer, that is precisely the locus of genuine homiletic power, a power that is real only insofar as it is realized by others. Can we say this of poetic power, or does a poem stand only within itself, and stand wholly autonomously? Is this a point at which we can truly distinguish poetry and prose, prose standing autonomously and poetry incapable of remaining within itself, or of standing alone? We sense that there is a deep difference between poetry and prose, one that we can sense all too literally, and actually know in our evocations of each, as our evocations of poetry unlike our evocations of prose can never be only our own.

This, too, is a challenge of poetry, and an all too intimate challenge, one that we actually seek, and seek insofar as we are speaking beings, and the only beings that can actually speak. How odd that philosophers have so ignored this, as though speaking does not fall within the province of philosophy, or not insofar as philosophy is an autonomous discipline. Is it only poetry that is the province of speech, or is so fully and comprehensively, so that a full expression of speech is inevitably a poetic expression? In this sense each of us are poets, or are so insofar as we are capable of genuine speech, and even if we can only know such speech as a gift, and a gift never within our own power, it nevertheless is realized only by ourselves. At no other point do we so realize the full conjunction of nature and grace, the deepest mystery of theology, and yet one actual in us in the actuality of our speech.

Augustine is the great theologian of nature and grace, that Augustine who created autobiography, and even created or first drew forth a uniquely Western self-consciousness. So, too, did Augustine create the doctrine of predestination, one based upon Paul's *Epistle to the Romans*, but uniquely Augustinian as a total doctrine, and one accepted and renewed by every major Western theologian, and at no other point is Western Christianity more distant from Eastern Christianity. Post-Classical Western poetry is for the most part within or integral to an Augustinian tradition, and an Augustinian tradition of grace, one inseparable from predestination, but nevertheless in an integral relationship with nature or the creation. This is a deeply intimate grace, hence it is the province of poetry and art, and is indeed the actual source of that poetry and art.

Now Augustine is not a poet, although he is a great master of prose, and while there have been poets such as Dante and Milton who are great theologians, these differ from modern poets who are theologians of a very different kind, and different if only because they are inevitably profoundly shaped by a secular or atheistic world. Hence we cannot theologically understand modern poets as we do a Dante or a Milton, but this is not to say that they cannot be understood theologically, but rather to accept that they require a very different kind of theological understanding, and one alien to all given or established theology. Perhaps the oddest of all modern poets theologically is T. S. Eliot, whose own theology is vastly distant from his greatest poetry, even including his late great theological poetry, which embodies profound theological polarities, and polarities that make possible this very poetry.

Eliot is the greatest critic of all poets, and his criticism is often theological, even going against his own theology, unless his deeper theology is in his poetry rather than his prose. *The Wasteland*[6] is itself a profoundly theological work, just as it has been truly influential theologically, and it is perhaps most theological in its sounding rather than in its sense, and as we speak its primal words, we inevitably do so liturgically, thereby evoking a truly new liturgical theology. Yes, this is an atheistic theology, but it is a holy or sacred theology, too, a holy atheism that however anathema to the deeply Christian Eliot, is nonetheless in continuity with the radically Christian poetry of Dante and Milton. Despite himself, Eliot is a profound witness to a radical Christianity, one that is purely present in that Blake whom Eliot loathed, and perhaps loathed as a mirror but reverse image of himself.

The greatest irony of our poetic tradition is the Satan of Milton, and Blake is its most powerful homiletic enactment; only here are Blake and Milton profoundly conjoined, and it is just here that each of them is both most powerful and most deeply poetic. Nowhere is modern theology more innocent or more naïve than in its distance from Milton and Blake, and above all its distance from the Satan of Milton and Blake, who is far more proximate to Augustine's *City of God* than to any later theological work. Now it is remarkable that the Satan of the mature Blake is virtually unknown, yet here lies the greatest homiletic power in all of our poetry, and even if here Satan is God the Creator, this is in no sense a Gnostic Creator, but rather the Creator who becomes all in all only through an absolute and total fall.

We reenact that fall in responding to Blake's great epics, *Milton* and *Jerusalem*, a reenactment that is inevitably a homiletic reenactment, just as is our reenactment of the fall of *Paradise Lost*. Yet such reenactment is inevitably a reenactment of Satan's fall, a fall that is our absolute beginning, a beginning inseparable from a renewal of the epics of Blake and Milton, and one even recurring throughout the great body of modern poetry. Here is that absolute negativity that is so offensive to the world, a negativity that is the center of Hegel's philosophy, and of Nietzsche's philosophy, too, but one whose most comprehensive embodiment is in modern poetry, an embodiment that is finally a redemptive embodiment, but one only possible through an absolute and total fall.

This is the fall that is so comprehensively enacted in modern poetry, but at bottom it is a *felix culpa* or fortunate fall, a fall apart from which no actual redemption is possible. So, too, no actual humanity or world is possible apart from this fall, or none known to the Western world, a world inseparable from a total fall, as known throughout the Christian epic tradition, which is just why Joyce and modernity lie within this tradition. Now the great theologian of a total fall is Augustine himself, whom just at this point is the most influential theologian in modernity, a modernity only theologically understandable through Augustine and his descendents. A total redemption is only understandable through a total fall, and if here Christianity and Buddhism are fully coordinate with each other, this is a coordination that can be discovered throughout our greatest modern poetry.

Tom

Letter 62

To Friends (June 26, 2012)

Friends,

Do you have a memory of my first book, *Oriental Mysticism and Biblical Eschatology*?[1] For some strange reason I have begun to reread it and am truly shocked! While much embarrassed by the very simple writing here, I cannot deny the real power of its first chapter on Greek religion, and I do relish returning to that youth when I intended to be a Greek scholar, just as I love revisiting vicariously those great scholars who so influenced me then. Nonetheless I am dumbfounded both by the section on tragedy and by the section on Aristotle, here is a theological understanding that I somehow lost, and it is a genuine theological understanding that is truly unique. Even now I am tempted to bring together this exploration of tragedy and Aristotle but wonder if I could now equal my earlier power. At that time I was far from what became my own theological thinking but I sense in this early book an openness to an ultimate theological "other" that now escapes me.

Only a short time ago I wrote a new chapter on the tragic Trinity for the Trinity book,[2] and that is surely a lesser understanding of tragedy than the one in my first book, but could this initial understanding have realized a genuine theological expression? Perhaps it does in the section on Aristotle and now I simply cannot imagine how I could have written that, although it was much influenced by great Aristotle scholars. And such a conjunction of great scholarship and theological thinking surely has become alien to our world, and yes, alien to me as well. I then said that it is Becoming that is indubitably real for Aristotle, and the reality of Becoming necessitates the "withdrawal" of God, for God has no reality apart from Becoming, because God is the actuality that is realized in the world. In reading this I can't believe that I was once a McKeon[3] student, but I remember how important Aristotle was in my undergraduate education, and how absent in my graduate education.

Despite the power of Thomism in the modern world, has a genuine Aristotelian thinking had any real theological impact, and has scholasticism made a recovery of Aristotle impossible? I don't think that this occurs in Leo Strauss or in anyone else whom I can now think of. Now we know the ultimate power of Neoplatonism even in our world, but who could imagine an Aristotilianism today? Yet in my first book I was certainly drawn to it, and somehow this occurred largely on my own, yet soon this disappeared. What happened? And how could I totally ignore this in my memoir? Indeed, is it not possible that it was Aristotle who made possible for me an atheistic theology? Remember the profound hatred that Luther had of Aristotle, knowing him as a genuine atheist, even if Luther only had a wholly indirect knowledge of Aristotle. Now if it was Aristotle who was the deeper father of scholasticism, and an Aristotle only discovered by Islam, is this the true or deeper origin of modern atheism?

A most puzzled,
Tom

Letter 63

To Friends (June 28, 2012)

Friends,

As I contemplate the possibility of rewriting my first book, *Oriental Mysticism and Biblical Eschatology*, I attempt to return to the distant time over 50 years ago when I wrote the book. This was a time of fundamental innocence for me, but also the time of my fullest freedom, when I was most free of commitments and responsibilities, and most open to truly new voyages. I was blessed in my doctoral father Joachim Wach, to whose memory the book is dedicated, not only was he a great bibliographical resource but he gave me the greatest possible freedom and support. I was also fortunate to be teaching at Emory when I wrote the book,[1] an Emory that then had the most radical theological faculty in the country, and the only major philosophy department that was wholly free of the curse of analytic philosophy. When a public furor developed over my becoming widely known as a death of God theologian, the Emory administration fully and decisively supported me, although under the threat of losing all of the support of their major benefactor if they did not.

As I attempt to return to the time prior to working on this book there is little to remember apart from the overwhelming impact of Nietzsche, one occurring to me even while I was teaching myself Kierkegaard and Barth, who were then ignored in the Chicago Divinity School, impelling me to transfer from theology to the history of religions. Perhaps I am alone in being so deeply shaped by Kierkegaard and Nietzsche, whose revolutionary work somehow flowed together in my youthful thinking, but this did make possible a genuinely new theology, which is decisively entered in my first book. However, this book is not yet a fully radical theology, but it prepares the way for one, and forges decisive steps to its full advent. I should also mention that it was while working on this book that I suffered the crisis of rejection for the priesthood by the Episcopal Church, one occurring because a required and thorough psychiatric exam had reached the judgment that I was deeply mentally ill and would soon have to be hospitalized, and could not possibly serve in the priesthood, although the judging bishop ignored my student pastorship of St Mark's Church, then the only interracial Episcopal church in the diocese of Chicago.

Note that I had made myself neo-orthodox theologically, was a deeply committed Christian, had even explored the possibility of a monastic vocation in an Anglican Benedictine monastery, and was prepared to devote my life to the Episcopal priesthood, being wholly alienated from the Roman Catholic Church because at that time before the Second Vatican Council there was no theological freedom at all in the Roman Church, and this at a time when there was a chasm between biblical scholarship and theology. Wach, himself a devout Anglican, thought that I had no genuine priestly vocation, and could become a far more effective theologian as a layman, a judgment that I gradually and decisively accepted. At that time neo-orthodox theology dominated the theological world, but it ever increasingly became unreal to me, as it was so ultimately distant from the actualities of our world, and this in a truly comprehensive if not total sense.

Originally my academic ambition had been to become a Greek scholar, and as an undergraduate I did study in the University of Chicago's Classics Department, but it was soon manifest that I did not possess the linguistic ability for such a vocation, even if I was deeply excited by the world of Classical scholarship. I read widely in this world, and this did lead to the first chapter of the book on Greek religion, which frankly even awes me a bit today, not because of my writing, but because of the great scholarship it reflects. I must be one of the few theologians who became truly aware of the power of Classical paganism, and I return to it in the first chapter of *History as Apocalypse* on Greek sculpture; apart from Ray Hart I am aware of no theologian who is open to the depths of the Greek spirit, and this is alien even to those German theologians who underwent a major Classical education. But it was decisive for me in becoming aware of the ultimate distance between the Classical world and the biblical world, and realizing that religiously or theologically neither is deeper than the other, and both embody ultimate religious worlds.

We have come to know, and to decisively know, that apocalypse is the very center of the New Testament, and of the prophetic revolution as well, but we have virtually no theological understanding of apocalypse, which has been alien to theology throughout its history, although it has been overwhelmingly powerful both in the Western imagination and in modern philosophy. My first book was an initial attempt to enter the challenge of apocalypse, which is here dealt with both in the Old and the New Testaments, but here Jesus is the primary challenge, for theology has never been able genuinely to enter his proclamation and enactment of the Kingdom of God. Here my primary ally was Rudolf Otto, that great historian of religions whose *Kingdom of God and the Son of Man*[2] is the deepest theological exploration of this topic, and who gave us the image of dawning for the Kingdom of God that Jesus enacted. Of course, Otto is ignored by New Testament scholarship, and by theology as well, but he was the primal master of Wach, and a primal master for me as well.

Now Otto also wrote the best book comparing Eastern and Western mysticism, *Mysticism East and West*,[3] which is an in depth comparison of Shankara[4] and Eckhart,[5] and one eclipsing virtually every book in this field. While I didn't initially realize it, Otto is the great scholar on mysticism and eschatology, and through Otto one can enter a fundamentally new theology, and one more genuinely biblical than its counterparts. I did learn that Bultmann was profoundly hostile towards Otto, and regarded him as his greatest theological rival, for Otto is more profoundly threatening to Bultmann and to New Testament theology as a whole than is Barth. This becomes manifest in the shallow understanding of religion pervading neo-orthodox theology, even if this is not true of Barth, or of the early Barth. Theology has become fascinated by the early Barth, a fascination only made possible by a turning away from the dogmatic Barth, but theology remains closed to the depths of religion, hence continuing to remain closed to the history of religions, and to the great scholars of religion.

Perhaps it was my very training in the history of religions that most made possible my becoming a radical theologian, it is difficult to imagine how a Church theology of any kind could now become open to religious depth, and this position is reflected in my first book. While there is a great deal in this book on both Barth and Bultmann, the word "Church" does not even appear in the index, just as a common or popular Christianity is wholly ignored, and ignored if only because irrelevant to serious inquiry. I have often debated with fundamentalist pastors, and assaulted them for their refusal of Jesus, which wholly baffled them if only because of their ignorance of biblical criticism and scholarship. Already in this first book I was attempting to give witness to Jesus, and recognizing thereby that this would require an ultimate transformation of theology, which had become closed to everything that we can either critically know or deeply imagine about Jesus, for Jesus is the ultimate X or the ultimately Unknown in our world.

I had specialized in Mahayana Buddhist philosophy as a graduate student in the History of Religions, and while Wach suspended the linguistic requirements because of his awareness that my interest here was wholly theological, Buddhism did have an enormous impact upon me and I later became close to Buddhist masters such as Abe[6] and Nishitani.[7] The chapter on Buddhism in my first book is not one that I can be proud of, but it did open the way for my continual theological correlation of Buddhism and Christianity, and did make me what one theologian, Herbert Richardson,[8] has called the first Buddhist-Christian theologian. This is also the one point in my work where I am wholly given to an in depth purely religious understanding, and can approach Buddhism as being more purely and profoundly religious than Christianity, and thus an ultimate challenge to Christianity. This chapter can teach me even now how much more profound Buddhism is than is Christianity, and if only in the perspective of Buddhist philosophy we can realize the shallowness of Christian theology, which even refuses to think about its own deepest ground, and this despite its endless discussion of God.

My first book ends with a chapter entitled "Faith and Being," and this chapter above all the others demands a radical revision, but it ends with a section entitled "Faith and the Nothing." Here, is certainly my best early theological thinking, and perhaps my best theological thinking, even if it is much too ambitious, and perhaps most so in attempting a full theological encounter with Nihilism. Now I can recognize that this has been a constant theme throughout my work, but how fascinating that this should be so central in my first book, in my very initiation as a theologian! Despite the fact that I published a book entitled *Godhead and the Nothing*, I am not wholly confident that I have ever gotten beyond my first book, and perhaps that is a confession of failure. Nonetheless the last chapter of the first book is a failure, except perhaps for its final sentence: "For just as the Buddhist comes to know samsara as Nirvana, the Christian must come to know the Nothing as the hither side of God." Amen!

Tom

Letter 64

To Friends (July 18, 2012)

Friends,

I have just returned from another visit with the late dramas of Eugene O'Neill, the greatest dramas of the twentieth century, and despite common opinion the greatest religious dramas of our world. There has been virtually no theological investigation of O'Neill, and even if this is impossible from the perspective of all conservative theologies, it is not impossible for a radical theology that is open to absolutely new enactments of God. It would be difficult to identify a body of dramatic work that is more profoundly religious than O'Neill's, the problem is that this is a sacred drama defying all of our understanding of the sacred, it is not atheist in a common sense, but it is profoundly atheistic in a deeper sense, and allied with a Blake or a Nietzsche, rather than a Marx or a Comte.

Of course, this is a radical atheism occurring throughout our late modern literature, and throughout our late modern drama, too, but it commonly defies both philosophical and theological understanding, and above all philosophical and theological understanding of God. It is extraordinarily difficult to speak about O'Neill and God, not only does the word seldom appear in his language, but there is a profound void of God in his dramas that is unique. And as opposed to Beckett, this is not an active or an actual void, but rather an invisible and a silent one, even if it wholly consumes all of the major characters in his dramas. O'Neill's is perhaps the most realistic of all drama, there is simply nothing in European drama truly parallel to such realism, and if this is a distinctively American realism, it is perhaps an embodiment of a distinctively American atheism, an atheism inseparable from a common language and experience.

America is paradoxically the most religious and the most areligious of all modern nations, and nowhere is this more fully embodied than in the dramas of O'Neill, the most American of all dramas, and perhaps even thereby the ones most defying a critical understanding, which is a decisive sign of their uniqueness. Indeed, all forms of drama seem to be conjoined by O'Neill, yet he nonetheless created the purest drama, and above all so in his late plays, *The Iceman Cometh* and *Long Day's Journey into Night*. The latter play was O'Neill's greatest challenge, and not only because it is an O'Neill family drama, but because he wrote it while in the deepest pain, and in the deepest agony over his own tragic life. Yet O'Neill universalizes tragedy as does no other tragedian, here it is realized as our common life, and a common life in confrontation with God, even if God is unnamable in this confrontation.

More than virtually any other dramatist, O'Neill's plots are truly dark plots, plots realizing darkness itself, and doing so even in their most enlightening movements. Although unspoken as such, it is God who is the ultimate enactor of these plots, for it cannot be fate or destiny, a destiny that in O'Neill's greatest dramas is simply life or actuality itself, and life in its most immediate expressions. No, it is a deeper ground that enacts O'Neill's plots, one probably unconscious to himself, and one demanding an ultimate courage, a courage to will darkness itself, and to will a darkness with no possibility at all of a reconciliation or atonement. Here is the dark God of O'Neill, one enacted throughout his dramas, but always as a nameless and hidden God, and above all hidden to O'Neill's characters. Yet it is precisely thereby that this God is most powerful, and most omnipresent, being most present in His very absence.

While redemption is wholly alien to O'Neill, damnation is continually enacted, a damnation only possible through the total absence of redemption, and a damnation that is enacted in virtually every play of O'Neill's, and far more here than in the work of any other playwright other than Beckett. There is a very odd coincidence between Beckett and O'Neill, they are totally different as playwrights, and yet totally alike in so centering upon damnation. Each is deeply Irish and yet in deep rebellion against Ireland, for O'Neill this is primarily a rebellion against the Catholic Church, one that is a genuine repetition compulsion in the Freudian sense, for it is a rebellion that can never cease. But it is a primal source of the dark plots of O'Neill, plots enacting a dark but hidden God, hence continually enacting damnation.

Could that damnation be a primal source of the unique realism of the dramas of O'Neill? This damnation is so real in these dramas as to make them wholly problematic as sacred dramas, but sacred dramas they are, and precisely by way of the damnation that they enact, for this could only be an eternal damnation, that very damnation that was so real in the Church that O'Neill knew. True, this is a Church that has wholly retreated, and no greater transformation has occurred of the Catholic Church, and if Catholic theologians no longer speak of damnation, Catholic artists do, and above all the greater Catholic artists, and if only at this point O'Neill is a Catholic artist. Without doubt, this is the point at which O'Neill can most spontaneously be entered theologically, but nothing is more alien to contemporary theology than damnation, so that O'Neill has been wholly closed to us theologically.

This is a genuine theological irony, for the God of damnation is deeply and continually enacted in O'Neill's dramas, and so much so as wholly to suspend every other epiphany or manifestation of God, a silence of God making possible that total damnation that is enacted in these dramas. Does such a silence of God make possible the profound realism of O'Neill's dramas, a truly unique realism, and unique as a purely dramatic realism? When participating in an O'Neill drama it is simply impossible to doubt it, simply impossible not to be enacting it oneself, but that is a profoundly threatening experience, if only because this is inevitably an enactment of one's own damnation. Never are we allowed to speak of it as such, or even to think of it, a decisive sign of a repression ever accompanying these dramas, and a repression that is yet another signature of Eugene O'Neill.

Nonetheless this is finally a liberating repression, and liberating by way of the actual drama that occurs, a drama that is truly real, and enacted as such by the spectator of the drama. This is a catharsis going far beyond any that Aristotle could understand, if only because Aristotle is closed to damnation, and in being so closed is closed to an ultimate depth, a depth again and again enacted in the dramas of O'Neill. Thereby a profound darkness is enacted, but in being realistically enacted it can be sanctioned, and sanctioned as darkness itself, a sanctioning realized in a genuine enactment, even if that enactment is an enactment of damnation, as occurs so profoundly in the dramas of O'Neill. Here, it is drama itself that is a liberator, and above all so a realistic drama, a drama that we actually enact, and precisely thereby can know not redemption, but a hallowing of our very acts, or of those acts that are most real to us.

Tom

Letter 65

To Friends (November 9, 2012)

Dear Friends,

I find myself compulsively revising my little essay on revolution and God, no doubt taking it far too seriously, but something like a return is occurring here, a return to my work on epic, and perhaps again a movement into the theology of history. Hegel is at the center of this, of course, but do any Hegelian philosophies or theologies of history remain, or is this unique to my calling? I do remember that as an undergraduate at the University of Chicago my first awakening was to the theology of history, one occurring wholly independently, and with something like a genuine calling. Then Spengler[1] had an enormous impact upon me, a Spengler who has now passed into oblivion except for Heidegger's confession of a youthful love for Spengler, and what I found most exciting in Spengler was his integral drawing together of mathematics, architecture, theology, metaphysics, and art, in his understanding of Classical and Faustian or Western and Islamic culture.

This deeply affected me theologically as I became persuaded that genuine theology incorporates allied fields, thus finally impelling me upon a quest for the Christian epic, a quest that I alone have pursued, a solitude that continues to baffle me. But this quest does illuminate both Hegel and Nietzsche, allowing us to understand them as epic thinkers, and perhaps they are most allied in this, an alliance apparently impenetrable to everyone, and even perhaps impenetrable to that Derrida who so embodied it. I think of epic as the most elusive of all genres, there is certainly far less critical work upon it, and the truth is that it can only be understood from a universal perspective, if only because genuine epic is not only addressed to but itself incorporates a universal world, as is most clearly manifest in Dante and Milton. I never could discover a critical study of Milton and Dante, but they do exist of Milton and Blake, and of Dante and Joyce, and when great epic visions are related or enjoined, one discovers genuine explosions. These are explosions that theology is challenged to incorporate, and even if my expositions are here far too weak, at least they have occurred, and perhaps not without effect.

Am I now being called to one last theological battle? That is what I would like to think, and I have only truly been alive when at war, and this is not only an intellectual or conceptual conflict, but genuine war, and a war even to death, a death and an eternal death apart from which theology loses all depth. Fascinating conversations with philosophers and theologians have occurred in my retirement, and I wonder if any of them could be real apart from eternal death, a death that not only pervades all of our deeper literature, but our deeper thinking as well, as Edith Wyschogrod once taught me. Edith could conjoin Hegel and Heidegger by way of the ultimacy of their enactment of death, but in Hegel this is an enactment of the death of God, and even if he commonly disguises it, it bursts forth in virtually all of his greatest thinking, and it is just at this point that Hegel is most profoundly resisted. Now if it is true that Hegel's is the only genuine philosophy of history, or Hegel's and Schelling's together, is this inseparable from a genuinely theological ground, which Hegel and Schelling share above all other modern philosophers?

I have long thought that both Hegel and Schelling are most resisted because of their theological ground, and here both have been at war with a subsequent modernity, just as their genuine descendents are all true warriors, and this already beginning with Kierkegaard and Marx. Here, nothing is more ironic than a professor of philosophy or theology, and if Hegel seemingly sanctioned this, at no point has he created more offense, or so effectively disguised his own philosophy. I always felt genuine guilt at being a professor, and never did I gain more relief than in retiring from the academic world, and even if I had deep friends in this world, it is a world in which innocence of any kind is simply impossible. So it is that I have always been contemptuous of academic writing, and for that matter of academic life, surely a primal source of our most destructive fantasies, including our philosophical and theological fantasies, which demand a true iconoclasm.

Certainly I have attempted a theological iconoclasm, and perhaps I am most deeply called to that today, just as it could well be that such an iconoclasm is most deeply needed in confronting our void, a void that is inescapable wherever there is any life today. So if I am being called once again to a theology of history, nothing could be more alien to our world, and perhaps that is just what most impels me, or most calls me to our Void.

Again a solitary,
Tom

Letter 66

To Friends (February 15, 2013)

Friends,

Nothing is more elusive to us than freedom, a freedom that we revere but cannot understand, and above all cannot understand in its most challenging expressions, expressions that are most elusive in their very agent or agency, an agent finally incapable of the actuality of freedom. This has issued in theologies affirming that grace alone is the source of freedom, a freedom wholly impossible in our given condition of impotence, and not only impossible but illusory as well, and even if that illusion is universal, nothing else so decisively marks our fallen condition. Here, we are at the very center of Western theology, as created by Paul and Augustine, one that has profoundly affected both the medieval and the modern worlds, and at no other point is theology itself more universal. Yet this theology occurs in a world that can know itself as the arena wherein the birth of freedom occurred, a freedom known nowhere else in the world, and certainly not as a universal freedom. But if our freedom has an origin or birth, then so, too can it come to an end, an ending both called forth and enacted in many of the most powerful expressions of late modernity, as most dramatically and universally enacted in our uniquely modern totalitarianism.

Totalitarianism differs from all previous tyrannies in its very reversal of freedom, as it evolves the deepest imprisonment that has ever occurred, a unique signature of late modernity, and one occurring not only historically, but imaginatively as well, as witness our greatest literature. Nothing is more indicative of this than the realization of a new and total responsibility, one that perhaps dawns in Dostoyevsky, and if no modern writer is more beloved than Dostoyevsky, none more profoundly challenges us, and above all so in his very enactment of a total responsibility. That responsibility is inseparable from a profound freedom, the only actual freedom that we can know, for genuine freedom is inseparable from a full responsibility, just as the absence of responsibility is the absence of freedom. Here, Dostoyevsky is very close to his beloved Shakespeare, and here, too, we can observe an immense distance between Greek tragedy and modern tragedy, such freedom being absent from Greek tragedy, and absent here if only because of the primacy of destiny. Of course, destiny is real and even ultimate in both Shakespeare and Dostoyevsky, but unlike Classical destiny this is a destiny that is a willed destiny, as occurs so powerfully in the heroes and the heroines of both Dostoyevsky and Shakespeare. Will itself was not discovered until Augustine, hence there is no awareness of it in the Classical world, but will becomes overwhelming in the very advent of the modern world, as fully manifest in the new world of modern tragedy.

Now if Hamlet is our primal tragic hero, there is not an absence of the will in Hamlet, but far rather a profound ambiguity or ambivalence of the will, here that ultimate self-division discovered by Paul becomes all consuming, and an absolute self-negation is realized that paradoxically calls forth the fullest possible will. This is a will realizing an absolute responsibility, and an absolute responsibility for everything, so that it inevitably realizes a total guilt, but that guilt is inseparable from the full actuality of freedom. Already Paul knew this profoundly, as did Augustine, and the Augustinian tradition itself, our only tradition knowing the depths of guilt and the depths of freedom simultaneously, a tradition that is an essential ground of modern tragedy, and of modern freedom itself. It is just the absence of guilt that unveils the vacuity of our pseudo-freedoms, pseudo-freedoms that abound in modernity as they do nowhere else, pseudo-freedoms creating our deepest illusions of freedom, and if there is no greater illusion in late modernity, nor a more pervasive one, this is an illusion perhaps more than any other unveiling our actual or genuine condition. Does our very imprisonment generate the illusion of freedom, or the illusion of modern freedom, one impossible apart from an interior bondage that was not even discovered until Paul?

Hegel, the creator of a philosophy of absolute self-negation, is our deepest Pauline thinker, or our deepest since Augustine, and just as Augustine created the theology of history in his *City of God*, Hegel created the philosophy of history, a history that is a forwardly evolving history, and it evolves by way of a progressive realization of freedom. Here, our historical destiny is apocalypse itself, and Hegel as our first apocalyptic philosopher is even thereby our first philosopher of the death of God, a death of God that is our deeper destiny, and only that death historically realizes an absolute freedom. One of the invisible or all too subtle births in the *Phenomenology of Spirit* is the philosophy of history itself, one that is a consequence of the realization of the death of God, for only that death makes possible an absolutely autonomous history, one that had never been known or actual before. That is the history in which a truly new freedom is born, but this is a freedom entailing an absolute responsibility, or that total responsibility so deeply known by Dostoyevsky, a total responsibility crushing or breaking humanity, thus effecting the end or ending of history itself. Not only does Hegel profoundly know this ending, but his philosophy would be impossible apart from it, and so likewise would be impossible the greatest expressions of the late modern imagination.

There is a unique meaning of freedom and responsibility in Christianity and Islam, and that is that its exercise must inevitably culminate in either redemption or damnation, damnation for the vast majority and redemption for the elect. While modern Christianity has deeply attempted to disguise this in virtually all of its expressions, it cannot be denied that nothing else is clearer in the New

Testament itself, or in all of the classical expressions of Christianity until the age of the Enlightenment. The most influential of modern theologians, Karl Barth, has argued in perhaps the most powerful section of his *Church Dogmatics* (2/1),[1] that as a consequence of the redemption of Christ damnation is impossible for all others. Barth is virtually alone as a theologian in maintaining that only Christ himself suffered damnation, one making possible the actuality of the atonement effected by the Crucifixion, an atonement that Barth was the first theologian fully to understand. Hence this is an atonement through damnation itself, or though the reversal of damnation, but a reversal impossible apart from damnation, and impossible apart from the damnation of Christ. Already Dostoyevsky understood this profoundly, although he did not dare to speak it clearly or openly, but it is nonetheless all too clear in his most theological novel, *The Possessed* or *The Devils*, and here clear in his most enigmatic character, Kirilov, who realizes the deepest imitation of Christ in all of Dostoyevsky's work. That is an imitation that could only be a realization of damnation, a damnation that when it is most actual we share with Christ, and this is the deepest center of the communion of the saints.

No theologian dares to rise to the level of Dostoyevsky, nor to the level of Shakespeare before him, and above all so no theologian within the Church, or within any sanctioned authority. How fortunate it is that we know so little of Shakespeare himself, and if we can know Shakespeare only through his work, and this work is so overwhelmed with commentary that it is virtually invisible by itself, that invisibility gives us a radical freedom in interpreting Shakespeare, but if we exercise it with responsibility we can actually enter truly new worlds. One of these is damnation itself, which is of so little interest to the commentators, but it is obviously a major motif of most of the great tragedies, just as Hamlet, Macbeth, Antony and Cleopatra, Otello, and Lear are not understandable apart from it. How shallow are the interpretations of the soliloquies of these great tragedies when they are interpreted wholly apart from all sense of damnation, just as these tragedies then become unreal as tragedy itself, for the great tragedies of Shakespeare are meaningless apart from damnation, and perhaps the great comedies as well if we could only understand them. It is the actuality of damnation that gives the dramas of Shakespeare a depth that is absent from Greek drama, for even Greek tragedy is closed to an ultimate horror, a horror returning again and again in the great tragedies of Shakespeare. Thus there is a depth of a *horror religiosus* that is virtually unique to Shakespeare, and one not truly paralleled by the Satan and Hell of *Paradise Lost*, and if Shakespeare is wholly alien to everything that William James knew as the once-born or the healthy-minded, this is not a Blakean Innocence, but far rather a Blakean Experience.

It is truly remarkable that the most famous of all soliloquies, "To be or not to be" (*Hamlet*, 3.1.64), is so commonly interpreted by way of a dissolution of all sense of damnation, as though this were not the very center of this soliloquy, but in being that center this gives an ultimate witness to the universality of damnation, even when damnation is wholly disguised. And we do disguise damnation, and disguise it as Shakespeare does not, and this is not simply because of the time and world in which Shakespeare lived, if only because it occurs so deeply in a uniquely Shakespearean language. Hence its primacy in the uniquely Shakespearean soliloquy, and if only at this point the Shakespearean soliloquy is unique, embodying a pure *horror religiosus*, as so forcefully evoked in "To be or not to be," which it should be impossible to hear without hearing damnation. Have we moderns chosen, if only unconsciously, to read and hear without hearing damnation, and even if this is clearly a distortion of Shakespeare, is that distortion necessary to the very survival of Shakespeare, to say nothing of his ultimately sacred status?

One of the supreme ironies of modernity is our uniquely modern forgetting of damnation, a damnation that is so primal in our most sacred literature, beginning with the New Testament, and culminating in a Joyce or a Beckett, and a damnation that is the most impenetrable of all possible philosophical topics, thus making philosophy irrelevant for so many. Yet there is one philosopher who has a profound understanding of damnation, and that is Nietzsche, and if only for this reason many refuse him as a philosopher, but he can be understood as being most purely philosophical in his understanding of damnation, or most purely original. A Nietzschean category for damnation is an absolute No-saying and No-saying alone, or a pure *ressentiment*, a total withdrawal from body that is an inversion of body itself, and hence a reversal of resurrection. Commonly it is said that we cannot understand resurrection, but we can understand its reversal, which fully occurs in the philosophy of Nietzsche, and above all so in his understanding of Eternal Recurrence. A Nietzschean Eternal Recurrence is the very opposite of a pagan eternal return, for it is an absolute affirmation or absolute Yes-saying, one that is the only possible source of an actual freedom, and the very incarnation of an absolute or total responsibility.

This is just the point at which Nietzsche is our greatest ethical philosopher since Spinoza, and if Nietzsche and Spinoza are our greatest ethical philosophers, this is clearest in their understanding of freedom and responsibility. Of course, each is a great opponent of freedom, but this is an opposition to a false or illusory freedom, or that very bondage that is commonly known or named as freedom. Apparently no real study has ever been done of Shakespeare and Nietzsche, but they share a great deal in their enactments and disenactments of freedom, and in the perspective of Nietzsche we can understand how ultimately radical Shakespeare is. This is clearest in his profound disenactments of what Hegel knows as the Given, that world or totality that happens to be at hand, or is simply given to us, a given that is apparently unchallengeable and inescapable. That is a given that Shakespearean language profoundly challenges, and challenges by wholly transfiguring the given, for it transfigures the given by resaying it in a language far beyond itself, but nonetheless a language in continuity with itself, a continuity making possible an actual and total transfiguration. This is the ultimate miracle of Shakespearean language, one most concretely evoked in *The Tempest*, as fantasy and reality pass into one another, and all difference melts away, or all difference that is not a linguistic difference.

Hegel is perhaps our best way into Shakespeare, a Hegel knowing and realizing totality itself far more than any other thinker, a totality realized in poetry and drama by Shakespeare alone, so if only thereby Shakespeare and Hegel are truly parallel figures. Only Hegel has dialectically identified an absolute necessity and an absolute freedom, thereby becoming our primary philosopher of freedom, but only insofar as he is our primary philosopher of a total responsibility. Hence Hegel made possible both Kierkegaard and Marx, thereby making possible not only dialectical materialism but the depths of modern freedom, and depths that can only be realized by a total responsibility. Of course, all of this if not foreseen by Shakespeare is actually enacted by Shakespeare, so that Marx and Kierkegaard loved Shakespeare as their primary source, and he is the primary source of modernity itself. Perhaps Shakespeare alone among all great artists is thought to be beyond both philosophy and theology, and beyond criticism as well, or any possible negative criticism, and if therein he is the very opposite of Hegel, this is yet another way of integrally relating Hegel and Shakespeare.

There is a vast range of Shakespearean character, far more so than in any other artist, but this is nonetheless a Shakespearean cosmos, and above all so as a linguistic cosmos. Nothing is more uniquely modern than are the characters realized by Shakespeare, at no other point is there such an immense distance between modernity and all other worlds, including most certainly our own world. In the perspective of Shakespeare we can realize just how monotonous our world is, and not only monotonous but anonymous, finally more anonymous than any other world, as we can realize in confronting the vast distance between our world and Shakespeare's world, and realizing that despite this distance our world is a descendent of Shakespeare's world, even if a very distant descendent.

Tom

Letter 67

To Friends (June 7, 2013)

Friends,

I have just finished reading the articles on Merleau-Ponty, Derrida, Deleuze, Badiou, and Althusser in the Stanford Encyclopedia of Philosophy[1], and while impressed with the authority and comprehensiveness of these articles, I missed in them anything truly fundamental or truly new. This may be simply because this encyclopedia arises out of a scholarly analytic philosophy, a philosophy closed to both the fundamental and the new, or it could be because twentieth century Continental philosophy apart from Husserl and Heidegger is so closed. Here there is a true gulf between German Idealism and contemporary philosophy, and even if a few have attempted to cross it, apparently our established contemporary philosophy has not. Now there is a parallel situation in a far weaker theology, one that I have been attempting to meet. This occurs through my commitment to Hegel and Nietzsche, whom I believe met the modern crisis of theology by recreating theology itself, for even if Nietzsche is our most anti-theological thinker, both he and Hegel recreated theology by profoundly and comprehensively reversing it.

Kierkegaard, too, recreated theology, and recreated it as a profoundly paradoxical theology, one issuing from an absolutely new discovery of the Incarnation, an incarnation creating a wholly new subjectivity, which Hegel knows as an absolutely new self-consciousness. Kierkegaard is our greatest modern theological thinker, who as such initiated Heidegger's theological thinking, as so profoundly embodied in *Being and Time*, even if that work is still not understood theologically. Christian apologetics has long taught that a primal way to an awakening of faith is a drawing forth of the deepest possible guilt, this surely occurs in *Being and Time*, but in a wholly disguised form, and here the impact of Kierkegaard is deep, even if this, too, is artfully disguised. We now know that the young Heidegger was profoundly affected by both Luther and Kierkegaard, and that he was deeply in quest of an original or primordial Christian faith that had been wholly transformed by the Church, and while the mature Heidegger is seemingly beyond this, he never ceases to be profoundly theological. Apart from Levinas this would appear to be true of no other established philosopher in our world, but just as Heidegger's theological quest was always deeply disguised, this could be true in the philosophical world at large today, and this might well be a disguise that is unrecognized by our philosophers themselves.

A genuinely contemporary theology could be most powerful when it is most disguised, and while we can now see this in Heidegger, could this be true wherever deep thinking occurs? Indeed, is not the history of philosophy the history of philosophical theology, or is so until Nietzsche? Nietzsche himself could know all philosophy as being grounded in God, which is just why his proclamation of the death of God is so revolutionary, and while Nietzsche was ignorant of the realization of the death of God in the *Phenomenology of Spirit*, he knew a depth of God both in our history and our consciousness that was closed to Hegel. It is just that depth that is the ultimate source of *ressentiment*, a *ressentiment* that Nietzsche discovered, but so likewise Nietzsche's revolutionary discovery of genealogy was the discovery of an absolute negation of God, occurring wherever there is an advent of the genuinely new, as most manifest in the advent of consciousness itself. Nietzsche can be known as the most powerful of all preachers, even if his deepest preaching is his proclamation of the death of God, but this is a genuine preaching, and one even having a kerygmatic effect, as manifest in all those who have been transformed by it.

If only in this perspective we can understand Nietzsche as a theologian, but Hegel is openly theological in his greatest thinking, and even the *Science of Logic* can be understood as an absolute self-emptying that is a self-emptying of Godhead itself, and one inseparable from its ultimate power. It is the overwhelming power of Hegel and Nietzsche that is a mystery to many, but this power is undeniable in our deeper contemporary philosophy, which would be inconceivable apart from both Hegel and Nietzsche. Yet that impact is inevitably a theological impact even if invisible as such, but thereby the deeper power of our contemporary philosophy is disguised, one disguising all ultimate depth, which is an inevitable consequence of disguising theology. One of the fascinating dimensions of the history of Western philosophy is its progressive silencing of God, and if this is reversed by German idealism, that idealism is a far more profound atheism, and one culminating in Nietzsche. This would appear to be an inescapable condition of contemporary philosophy, and if that philosophy beginning with Heidegger refuses all pronunciation of God, that could nonetheless be a deep disguise of God, as it surely is in Heidegger himself.

Is our philosophy in spite of itself nonetheless a theology? Clearly we can ask this of Heidegger, but can we ask it of every philosophy in our world, or of every philosophy embodying depth? Perhaps nothing so embodies the deeper history of theology than does the Christian epic, and no greater transformation of theology occurs than here occurs between Dante and Joyce, just as nothing is more purely theological than the epic enactments of Milton and Blake. The death of God is first enacted by Blake, in his first epic, *America*, and just as Blake is a genuine twin of Nietzsche, there is a genuine parallel between the history of the Christian epic and the history of modern philosophy. There is a progressive disappearance of God in each of these worlds, and Joyce can illuminate contemporary philosophy as perhaps no philosopher can, or at least illuminate it theologically. There is a progressive disappearance of God in *Ulysses* and *Finnegans Wake,* which is nonetheless a resurrection of God, a resurrection wholly different from the resurrection of the Christian tradition, and one occurring even now. This occurs through a renewal of the Eucharist, and the *Wake* is the only literary work that is an enactment and a pure enactment of the Eucharist, even if an enactment reversing the Catholic Eucharist.[2] Thereby the Eucharist itself is resurrected, resurrected from death itself, and here resurrected from the death of God, here Crucifixion *is* Resurrection, just as it is in Paul, or in the original Paul. Is there anything comparable to this in philosophy? There certainly is in Hegel, and perhaps in Nietzsche, too, but can resurrection be discovered in contemporary philosophy?

That is perhaps the supreme theological question that can be asked of contemporary philosophy, but it should be noted that nothing is more elusive or controversial in theology today than is resurrection, just as nothing else has been so paganized in the Christian tradition. Although unknown to most Christians, immortality is truly alien to the Bible, and the resurrection of the New Testament is wholly other than a pagan immortality. Resurrection has been gloriously enacted poetically, as occurs throughout the Christian epic, but it has not been understood theologically, or decisively understood, even if it has been deeply understood by Spinoza and Hegel, and perhaps by Nietzsche too. Here lies an all too ironical difference between philosophy and theology, with at least certain expressions of philosophy decisively understanding a primal theological category, as theology itself cannot, perhaps because theology has been closed to a revolutionary thinking. Or can we here affirm that theology is most purely realized not in itself but in the deeper expressions of philosophy, and even if these are all purely heterodox in the perspective of the Church, could we say that this is only true of the nonapocalyptic Church, or of that Church that is closed to apocalypse?

Tom

Letter 68

To Friends (August 6, 2014)

Friends,

Last night I resaw Penderecki's great opera, *The Devils of Loudon*,[1] in my opinion the most important contemporary opera, one that has been given a glorious recording on DVD.[2] No doubt I now returned to it because I am so seriously considering doing a book on Radical Catholicism, and unfortunately none of my writing in this arena examines this or any opera. You may remember the enormous impact that Penderecki made with his early St. Luke Passion,[3] the most violent of all passions, but nonetheless a profoundly Christian passion. That passion is renewed in *Devils*, as now the radically saintly priest of Loudon undergoes crucifixion in a world wholly dominated by demons, and a world in which it is here proclaimed that God is dead. Is it only here that these primal words are operatically enacted, and enacted in a truly sacred drama, even if a reverse sacred drama enacting the ultimate death of God?

Although this work was affected by Aldous Huxley's reenactment of Loudon, it goes far beyond Huxley or any other critic, and does so with a profundity rivaling Schoenberg's *Moses und Aaron*,[4] as a *horror religiosus* is released in each, and a genuinely apocalyptic horror. Unfortunately our critics are almost all incapable of entering an apocalyptic realm, hence an apocalyptic opera is unheard of, but it is overwhelmingly enacted in *Devils*, as an apocalyptic epic is here realized operatically. The parallel with Wagner is quite obvious, but *Devils* is a deeply Christian opera, issuing from a profoundly Christian composer, and why is it not acknowledged that late modern music is the most deeply religious of all the arts in this world? In *Devils* the religious depth is a purely Christian one, indeed, a profoundly Catholic one, but this is Radical Catholicism, and as purely a Radical Catholicism as we have been given.

One has only to experience a great production of this opera to experience and to know Radical Catholicism, but is Radical Catholicism itself deeply offensive, and turned away from everywhere? The Wikipedia article on Penderecki does not mention this opera,[5] but it does mention a number of major films employing the music of Penderecki, and we can certainly understand him as a truly major artist, and no doubt a profoundly Catholic artist. Our critics seldom explore the religious dimension of ours or any music, and while admittedly this is very difficult, it is essential to understanding this music, a music much of which is as religious as any other music, but a music that is seemingly untouchable theologically, perhaps because it is not only radical musically, but radical theologically as well. Here there is an integral relationship between radical art and Radical Catholicism, one paralleling this relationship in Dante and Joyce, and just as a theological entrance into Dante and Joyce is overwhelmingly difficult, that is true of Penderecki, too.

A deep problem here is the virtual invisibility of radical theology in our world, and above all the invisibility of a radical Catholic theology, a void that must be filled by non-Catholic theologians. Yet the truth is that our great Catholic artists create their own theology, as so decisively manifest in Dante and Joyce, and at no point is our Dante and Joyce scholarship so weak as it is in the theological arena. Nothing is more banned in our world than theology itself, and perhaps nothing so cripples our interpretation of our music as does this ban, a problem that is overwhelming in a profoundly religious composer such as Penderecki. Yet this is a very radical religion, indeed, and a profoundly shocking one, and above all in *Devils*, as Dostoyevsky here appears to be reborn, a rebirth above all of his *Devils*.

While Penderecki is deeply Catholic, he is also deeply Polish, and hence not alien to the world of Dostoyevsky, but each of these worlds are theologically unknown, and that poses a profound hermeneutical problem. This is most manifest in the very conjunction of radical music and radical religion, and while we have an understanding of radical music we have very little understanding of radical religion, so that we are wholly baffled by an opera like *Devils*. But so too are we baffled by Wagner, for despite the enormous literature on Wagner there is apparently no responsible theological interpretation of Wagner, a problem compounded by the enormous distance of Wagnerian opera from all of our established theology.

Nietzsche's late revulsion against Wagner is revealing, one only made possible by his early passion for Wagner, a genuinely religious passion, and one that exploded when Nietzsche created an ultimately radical theological thinking. Nietzsche could respond to *Parsifal*[6] as an ultimate assault upon his own world,[7] a *Parsifal* that is the most religious of Wagner's operas, and even the most Christian, which was above all offensive to Nietzsche. Can we imagine a reaction of Nietzsche to *The Devils of Loudon*? Would it not be a total Yes, and this not only because this is the fullest operatic enactment of the death of God, but also the fullest operatic enactment of an absolutely pathological religious passion, one that Nietzsche profoundly discovered? However, Nietzsche could only be unresponsive to the Catholic ground of *Devils*, and it is difficult to think of one who could respond to this, if only because Radical Catholicism is so deeply unknown. Yet simply by being aware of how alien Joyce and Dante are to a given or established Catholicism, we can become open to the Catholic ground of, one only possible as a Radical Catholicism, and thereby demonstrating the reality of Radical Catholicism itself.

A demonic Tom

Letter 69

To Friends (September 22, 2014)

Friends,

Our most revolutionary prophet, William Blake, in his first prophetic poem, *America* (1793), enacted the American Revolution as the initial realization of the death of God, the deity here named as Urizon, the preincarnate and alien God, whose death initiates apocalypse. This is the God whom Hegel named as Abstract Spirit and the "Bad Infinite," a God not realized until the advent of the modern world, and who is the consequence of an absolute self-negation or self-emptying of the Godhead. Both Blake and Hegel enact the death of God, indeed Hegel and Blake are the first enactors of the death of God, a death that for each is an absolute self-negation or self-emptying, a self-negation that is the absolute source of all and everything. Hence the death of God is both genesis and apocalypse, or absolute beginning and absolute ending, the absolute beginning of all and everything, and the absolute consummation of everything. That consummation itself proceeds out of an original self-negation or self-emptying, one negating or emptying an original absolutely undifferentiated Godhead, and only this self-negation makes possible apocalypse itself. Hegel is our most profoundly apocalyptic thinker, while Blake is our most totally apocalyptic visionary, each recover and renew a long lost apocalyptic ground, a ground that is the original ground of Christianity, one that is wholly transformed in the great body of Christianity, and only recovered in revolutionary movements, which are the most revolutionary movements in our history.

Both Blake and Hegel are profoundly Christian, but they are radical Christians, even atheistic Christians, who absolutely negate the given God, or who deeply and comprehensively realize that this God has absolutely negated itself, a self-negation that inaugurates the modern world. Each could know the French Revolution as the historical realization of the death of God, but Blake, at least in *America*, could know the American Revolution not only as the initial realization of the death of God, but as the inaugurator of absolute revolution. This is the deepest calling of America, one known to every deeply American seer, and actualized in that America that is the first secular nation, the first not only to separate Church and State, but to create a public realm that is a truly secular realm. This inspired an assault upon America by many European Christians, but Europeans have never been able to understand America, and the question can be genuinely asked if America has ever understood itself.

America, with the possible exception of Indonesia[1], is the most pluralistic nation in the world[2], the one with the largest number of diverse minorities, and whose original aristocracy was the first aristocracy to be wholly transformed, and now only pale traces of it remain. Of course, America is the most Capitalistic nation, now serving as a model for a once revolutionary China, but American Capitalism is now inseparable from American Militarism, and if American military power now transcends world military power, this power is having an immense impact upon America, where no political leader dares resist it. Once it was common to look upon America as the new Rome, and while the American empire is not as powerful as was the Roman empire, the impact of America upon the world is possibly as great as was Rome's, and not only in the pragmatic arena, but in a new and wholly new mass popular culture. America is the creator of that culture, and here one can see an all too concrete realization of the death of God, as depth itself is ended, or is now indistinguishable from surface, and just as American folk music and folk art are now disappearing, that is occurring throughout the world.

Innumerable thinkers have attempted to contend with the advent of a new mass culture and mass society, one that dawns in America and is rapidly extended throughout the world, and if only here America dominates the world. But is that culture inseparable from a vast economic and military power, so that the American empire is a truly new empire, as is fully manifest in the perspective of the Roman Empire, for Roman military power was a purely naked power, whereas American military power is disguised by its conjunction with our mass culture, and inevitably wholly disguised in the mass media. So in a fundamental sense American military power is invisible, but all the more powerful just because of that, for thereby it does not provoke ultimate resistance, and this despite the vast sums that are spent upon it, as a military totality has been created even in the absence of a military power or powers that could possibly contest it. Now this is very much like ancient Rome, and if that Rome had resisted imperialistic ventures it could possibly have survived, but can a genuine empire resist imperialism, which is certainly a primary question being asked of America today.

There is also a parallel between Rome and America in the religious arena, imperial Rome was the site of a large number of dynamic religious bodies including Christianity, just as America is the site of such bodies, and in Rome these bodies underwent enormous transformations, and most of all Christianity itself, undergoing a total transformation unparalled by any other religion in the world. Now Catholicism truly becomes Roman Catholicism, but is a comparable transformation occurring in our world, and could this be occurring in the context of the death of an old religion, as occurred in ancient Rome? Surely the death of God in the modern world is comparable to the twilight of the gods in the ancient world, Wagner realized this profoundly, but if America is the original site of the death of God, could that be essential to American destiny, even if that is now most hidden in that destiny? America is perhaps the most religious nation in the industrial world, but we know all too little about this, for whereas deep and comprehensive studies occur of the American economy, there is very little study of contemporary religious life, and no in depth studies of it at all, it is as though an absolute prohibition of inquiry has been enforced here.

American literature is perhaps the arena in which such an inquiry could now most decisively occur, and it is vitally important to attend to what is most unique here, what is vitally and most distinctively American. *Moby-Dick* is an apt site for such an investigation, for it is our most American epic, and also the first mythical novel, a truly fundamental breakthrough, and if the White Whale is the most powerful figure in American literature, it is truly a *horror religiosus,* and one profoundly enacting the death of God. Moby Dick is truly a Urizen in the American world, and a Urizen perhaps here undergoing an ultimate and final death, but is that really possible in the context of this novel, or in the context of America itself? Now if *Moby-Dick* truly is the American epic, is this an epic revolving about the death of God, a death of God that Blake could enact as the origin of America, and one dominating American literature as it does no other literature? It is surely paradoxical that such a seemingly conservative nation as America could produce such a radical literature, and this is true of its greatest literature, as is true of no other nation.

Eugene O'Neill is commonly accepted as the greatest American dramatist, and his major plays are enactments of the death of God in numerous arenas, plays that can be understood as our most nihilistic dramas, and yet they are deeply and even uniquely American. O'Neill is genuinely a revolutionary dramatist, perhaps the most revolutionary of all dramatists apart from Shakespeare, yet he is a Catholic dramatist, but only as a radical Catholic, and a radical Catholic opening himself to our underworld, creating an American inferno comprehending Here Comes Everybody[3], as most purely occurring in *The Iceman Cometh.*[4] But this play can finally be understood as a purgatory rather than an inferno, even if a nihilistic purgatory, which is perhaps the only purgatory that could be real in a genuinely American world. Certainly O'Neill is deeply American, and a European intending to study America, could do no better than to study O'Neill. But is O'Neill truly representative of American literature as a whole, or of a deeper American literature?

Insofar as we accept *Moby-Dick* as the American epic, we can become open to a distinctively American wildness or wilderness, a wilderness that deeply shaped America, giving it a uniquely American identity, an identity that is continually renewed in American literature. So it is that the fiction of Henry James, perhaps the most sophisticated of all fiction, could be understood as a sublimation of an American wilderness, but one only possible by way of that wilderness. James Baldwin, another sophisticated American novelist, once declared that the Gothic cathedrals are not part of *his* past[5], and this could be said by virtually any American, for America is truly a desert insofar as it is so isolated from the actualities of history. So it is that American Gothic is a truly ironic American architecture, one wholly artificial and false as a Gothic architecture, but one despite this fully pragmatic, and thus fully functional. Accordingly, Heidegger could reach the judgment in the conclusion of his *Introduction to Metaphysics*[6] that America and Soviet Russia are metaphysically identical, for here time as history as vanished from human life.

Nothing is more revealing about America than a distinctively American religion, and if America is the most religious of all industrial nations, its religious life is more diverse than that of any other nation, for if only because of the immense emigration to America, virtually every religion in the world has found a home in America. Not only did America initiate the separation of Church and State, it inaugurated the first truly secular culture, and only in America has religion existed in an integral relation with secular culture, thus leading not only to deep conflict, but to a deep religious absorption of secular culture, or a distinctly American idolatry. Such an idolatry can be observed in that mass advertising created by America, one in which wholly empty and vacuous images create a totally empty language, but one truly effective in that very emptiness, one creating empty but all consuming desire. There is something deeply American in mass advertising, a peculiarly American pragmatic power, one either truly demonic or wholly amoral, and one that could only occur in a deeply anonymous world, a nameless world that itself is peculiarly American.

Jazz is commonly judged to be the most distinctively American art, and jazz has a deeply religious origin in spirituals and the blues, blues that are truly unique to America, one conjoining the impact of slavery and oppression. Nonetheless there is a deep joy in the blues, one deeply realized in jazz, and openly embodied in the great blues and jazz singers, singers who have had a universal impact, and a universal liberating impact. But this is a very strange liberation, one seemingly suppressing or dissolving all possibility of rebellion and revolt, as though jazz is the source of a true passivity, but one nevertheless inseparable from a genuine joy. While America claims to be an ultimate source of freedom, it has been more deeply scarred by slavery than any other nation, thereby incurring a peculiarly American guilt, and one as deep as anything else in the American soul. Is there a distinctively American atonement or purgation for this guilt, one deeply occurring in American literature, as perhaps most manifestly realized in the major novels of Faulkner, that deep son of the state most embroiled in slavery[7], and yet a state that produced more major literature than any other state?[8]

Jazz itself can be thought of as a primal source of atonement, an atonement realized through the violence of jazz, a violence assuaging itself in its own actualization, as its assault upon us induces a deep calm, but a calm inseparable from assault. A pure expression of this is "The West End Blues"[9] of Louis Armstrong[10], one creating a truly new sound, assaulting all that is simply given as jazz, and transfiguring it into a new ecstasy, a pure ecstasy consuming its hearer, but doing so only by effecting an actual purgation, and an actual purgation that can be understood as an atonement, and an actual atonement releasing us into a new life. Jazz is inseparable from festival, a festival that is inevitably a sacred festival, as are all the high moments of jazz, but thereby and therein these are atoning moments, moments impossible apart from an actual purgation. At this crucial point we can only speak of atonement itself, an atonement resisting all real theological understanding, yet an atonement that has given us perhaps our greatest art, and if only here painting and music are in full harmony with each other, a harmony that is echoed in jazz.

If we associate jazz with the death of God, and do so because jazz is so deeply American, then the death of God can be understood as realizing an ultimate joy, one that Nietzsche himself profoundly proclaimed, and did so while himself in a wholly broken condition. For the joy released by the death of God is wholly independent of the condition of its recipient, or can only be correlated with a broken condition, that very condition that Nietzsche associated with a slave morality, a morality only possible by way of the reversal of a noble morality. Is jazz only possible through a reversal of a noble morality, does its very advent depend upon an ultimate assault upon the high and the exalted, one shattering all sanctioned authority and status? Many conservatives associate such an assault with a new America, an America transcending an established law and order, and inaugurating the first anarchistic nation. While such an America cannot be discovered in a common or established America, it is present in a radical America, as manifest in much of the literature of America, and even in its more radical religious bodies.

This is the America that called Blake and many other radical Europeans, but an America that can only be recovered through rigorous historical study, or by an imaginative or religious rebirth, for this is an America that has become hidden from view, and truly silent apart from its underground expressions. Now that America is the most powerful nation in the world, it is thereby the most established of all nations, and the one most sanctioning all established authority and power, a sanctioning inevitably occurring through its own power, but an overwhelming power inevitably inducing rebellions against itself. Are such rebellions now occurring in America, and even occurring through a distinctively American realization of the death of God? Few have given attention to the question of a peculiarly American realization of the death of God, and perhaps the deepest inquiry into this question is in H. L. Mencken's great book on the American language[11], wherein he demonstrates an absence in that language of a virtually universal linguistic and hierarchical distinction between high and low, a hierarchy wholly alien to the American language, and perhaps finally alien to a deeper America.

Certainly the death of God, or an actual death of God, dissolves or destroys all such hierarchy, and even if few Americans are aware of this, genuine conservatives know it all too well, and if modern conservatism was founded by Burke's response to the French Revolution, modern conservatism is a deep reaction against a revolutionary Terror destroying all established law and order, a Terror that is an enactment of the death of God, as proclaimed on the banners of the revolutionary armies. Once the deepest political controversy in America was over the French Revolution, one giving birth to a distinctly American conservatism, and even if this has now been forgotten in America, American conservatism remains deeply bound to all established law and order, and precisely thereby is the most powerful opponent of all revolution. Little is more ironical in world history than the American Revolution, a revolution that many could identify as the very advent of freedom, but a revolution that many can now know as ending all possibility of revolution, and doing so with an apocalyptic finality.

Already the young Blake in *America* gave us an apocalyptic vision of the American Revolution, a Blake who was himself inspired by American revolutionaries, and even if that is inconceivable today, we can thereby understand how an ultimate reversal of America has indeed occurred. The visionary Blake actually envisioned how such an absolute reversal occurred in Christianity, one finally transforming the sovereignty of the Christian God into a Satanic sovereignty, as most decisively occurring in his illustrations to the Book of Job, but as most fully occurring in his penultimate epic, *Milton*. Thus that radical Milton who first fully envisioned Satan is truly reborn in Blake, and most reborn in Blake's *Milton*, an epic giving us our purest vision of Satan, a Satan who is an absolute and universal negativity. Yet that negativity undergoes its own reversal in what Blake envisioned as "The Self-Annihilation of God," a self-annihilation fully paralleling a Hegelian self-negation or self-emptying of God. And if this is an absolute atonement for both Blake and Hegel, it is an atonement that is a revolutionary atonement, and one absolutely transfiguring everything whatsoever.

Has America been given the promise of enacting such a revolutionary atonement, and could this occur through its very embodiment of the death of God, an embodiment already negatively or inversely enacted in *Moby Dick*? Is it really so very odd so to associate America and the death of God, and if we approach this question through a Blakean and Hegelian perspective, does that not give America an ultimate identity that otherwise is inexplicable? Of course, many would oppose all ultimate identity, and think it most ludicrous to give America an ultimate identity, that very America that is now a new Rome. But is not a new Rome an ultimate identity, and even an apocalyptic identity, it would surely be in continuity with the Book of Revelation, a book profoundly naming an old Rome, and a book perhaps harboring an apocalyptic naming of America? When one reflects upon the role that America has played in realizing a contemporary world of universal economic exploitation and a universal mass society and culture, it would not be odd to give America such a purely demonic identity.

America's darker history has largely been hidden from itself, as perhaps most dramatically occurring in the Mexican-American War of 1846–1848[12], a war initiated by the United States, a war that is surely the most successful war in history, for it resulted in the annexation of most of what was to become the territory of the United States, and this at little cost to itself.[13] Thereby America was born as an imperialistic nation, a seemingly painless birth, but one transforming the United States as nothing else has, even if this transformation is invisible to Americans themselves. If this war can be understood as the historical advent of a demonic America, nothing has been more disguised in the American consciousness than this advent, which is the advent of a demonic America. Yet apocalyptically a demonic identity is inseparable from a salvific identity, the kingdom of darkness is inseparable from the kingdom of light, and absolute darkness only fully dawns in the wake of the advent of absolute light. Thus if America has a genuinely demonic apocalyptic identity that identity is inseparable from its very opposite, and if Blake is our most revolutionary apocalyptic visionary, at no point is that vision more fully manifest than in his vision of Satan. For in giving us our fullest vision of Satan, that very vision enacts the absolute necessity of Satan, and the absolute necessity of Satan for an absolute apocalypse. Even if he couldn't fully face it, Nietzsche knew the absolute necessity of evil, a necessity for realizing an actual Eternal Recurrence, a necessity apart from which Eternal Recurrence could only be eternal return. Eternal Recurrence can be known as our purest vision of apocalypse, and Nietzsche is a purely apocalyptic thinker, at which point he is in genuine continuity with the apocalyptic Hegel. In both Nietzsche and Hegel, those thinkers who most profoundly enact the death of God, there is an ultimate renewal of Joachism, that Joachism that is the deepest and most influential of all medieval heterodoxies, and above all so in its witness to the actual advent of the final Age of the Spirit.

Can America truly be dissociated from such an advent? Surely its own visionaries have known it, as concretely manifest in the most American of all new religious movements, Mormonism and Christian Science, both of which enact an Age of the Spirit, even if in all too limited and narrow modes. So, too, Christian apocalyptic sects are apparently most powerful in America, and Christian apocalyptic visionaries appear again and again in America, and do so ever attracting new followers, for America more than any other nation is overwhelmed by Christian sects, and this is true even while American Catholic and Orthodox bodies are more vital than they are in any old world. Nothing so baffles the European as does the vitality of American religion, but if this very vitality can be associated with the death of God, then it can become understandable, and perhaps understandable even to the European, and as European theologians have noted, when God is dead, religion is everywhere. Certainly many Europeans are horrified by the demonic power of America, and if only as a demonic power American can be known as embodying the death of God, a death of God that may well be more universal in America than anywhere else, even if thereby most deeply disguised.

Let us return to the symbolic figure of Moby Dick, who can be understood as the deepest symbol of America, and here an America who actually kills *God*, even if that murder is at bottom a suicide, and a suicide that can be understood as either the death or the self-annihilation of God. *Moby-Dick* is a truly prophetic book, and one envisioning not only a future but a present actuality, an actuality occurring in a deeper America, and one inseparable from everything that is genuinely American. Just as an absolute horror is actual here, a horror that is a *horror religiosus*, this is a horror that cannot be disjoined from a deeper America, nor disjoined from a deeper American destiny, a destiny called forth in the deepest American vision. A comparable horror is enacted in the dramas of O'Neill, whose greatest plays are our most horrible dramas, and just as these are commonly and critically accepted as our most American dramas, they inevitably embody a uniquely American destiny, which is everything but the manifest destiny of the nineteenth century.

While there are those who think that tragedy is alien to America, the opposite would appear to be true in American literature itself, a literature astutely ignored by the populizers of America, for whom America is a pure innocence. Let us recall that the American Civil War was the first totally horrible war in history, and a war producing an all pervasive horror of war, one deeply and comprehensively transforming America. The horror of that war can be known as being renewed in the First Word War, a war initiating America as a truly world power, but a power soon destined to be engaged in futile and self-destructive wars. Subsequent to the Second World War, American wars have been truly ironic, their victories inevitably become defeats, just as while realizing the greatest military power in history, this power itself is inseparable from a deep American impotence that no one can understand. Why not employ the symbol of the death of God to explicate this condition? Cannot contemporary American power be likened to the dead or dying body of God that destroys everything in its wake?

Yes, the symbol of the death of God is our best symbol for explicating America, an America that is inevitably a tragic nation, and even is so in its very innocence, an innocence that is inevitably an innocence lost. Nowhere can this be observed more decisively than in American literature, perhaps the most confessional literature in the world, and one embodying more horror than any other literature. While American critics may be blind to this, European critics are not, and while we may be astounded that a Baudelaire[14] could take American literature so seriously, Baudelaire knew the death of God more profoundly than any American. Americans do know the death of God, as truly manifest in American literature, but they deeply veiled this from themselves until only recently, when the death of God exploded in America, even if this explosion was soon dissipated. That explosion was a fulfillment of a long American gestation, one occurring in American philosophy as well as in American literature, and even occurring in American historiography and sociology that are truly radical American ventures.

It is remarkable how America has so fully hidden its revolutionary identity from itself, while so commonly proclaiming a Utopian identity for itself, a Utopian identity that is the very opposite of a revolutionary identity, one wholly disguising American from itself. Yet an originally revolutionary America has become the most reactionary of all contemporary nations, and this is the context in which it can be liberating to understand America as an embodiment of the death of God, thereby identifying America as a revolutionary rather than a reactionary power. While the death of God can release profound and deadly reaction, as so deeply understood by Dostoyevsky, it also can be profoundly liberating, a liberation ultimately enacted by Nietzsche, and by Hegel, too. And if it is Hegel and Nietzsche who most profoundly understand a uniquely Western destiny, that destiny cannot be divorced from the death of God, and from the death of God as an ultimate and apocalyptic event.

While we commonly think of modernity as fully dawning in the seventeenth century, and doing so through genuine political, social, economic, literary, philosophical and scientific revolutions, it is actually not until the nineteenth century that there is a full awareness and embodiment of the actuality of these revolutions, and only then is there an absolute break between an old world and a new world. Thus an actual language of the death of God is now released for the first time, and only now is the death of God either poetically or philosophically enacted, as occurs in the revolutionary work of Blake and Hegel. This is just the point at which Blake and Hegel are most openly revolutionary, for they are the first actually to speak or write of the death of God, which each understands as the advent of an absolute transformation, and a transformation transforming everything whatsoever.

America can be understood as the primal site of that transformation, or the primal public site, and if only for this reason American must be deeply disguised from itself, and wear the mask of a simple innocence. How appropriate that the ultimately revolutionary Blake should choose America as the point at which to inaugurate his own prophetic poetry and designs, thereby truly reversing the public image of America, and even calling forth the American Revolution as the ending of monarchic empire. Blake never returns to this image of America, and if no American prophet even approaches the revolutionary power of Blake, an ultimate disguise may well be necessary for a uniquely American destiny, and that destiny is clearly a universal destiny, as actually enacted in the uniquely American philosophy of Pragmatism. Here is a thinking that is truly American and truly universal at once, and if this is paralleled in a uniquely American drama, that is a drama centered in the death of God more than any other drama apart from Brecht and Beckett.

Can a contemporary America realize a uniquely American destiny? This is even now occurring if we identify that destiny with a truly anonymous and empty body, a body of no return because it has already wholly arrived, and a body that American prophets have enacted as a Great Sleep. While American destiny, as Henry Miller envisioned it, may well be an air-conditioned nightmare[15], that could be a realization of world destiny, and itself an apocalyptic enactment of the death of God. Is it, then, impossible to dissociate America and the death of God? Is it only America that can be envisioned as totally enacting the death of God, and is that truly America's destiny, and America's unique destiny? Dare we say Yes to this? Or is it possible to say No, and to say No with an actual and genuine conviction?

Letter 70

To Friends (September 22, 2014)

Dear Friends,

Having just wrestled with the possibility of cancer, and hence wrestled with death, I now ask about the relationship between cancer and theology. Perhaps theology only transcends philosophy in the arena of death, a death that theology knows as Death, here Heidegger can most decisively be known as a theologian, just as Heidegger was deeply engaged with a seminar on Paul when he wrote *Being and Time*. Death is also that realm that most distinguishes Judaism and Christianity, death being secondary in the Hebrew Bible and primary in the New Testament. Just as death is primal in Christianity as it is in no other world tradition, so, too, it is only in the Christian West that death is a fundamental arena of the imagination, only here that epic itself revolves about death, and that death is a deep center both of drama and of literature itself, as though it is only the Western imagination that revolves about death.

Perhaps Milton and Blake are our greatest visionaries of death, and not insignificantly they are also our most primal prophetic visionaries, as here the ancient Hebrew prophets are reborn, and reborn as they never are in our theology. Here, Milton's blindness is of great significance just as is Blake's poverty, for their overwhelming prophetic power is inseparable from a profound distance from the world or the worldly as such, or from all power in the world. This is the perspective in which we can think of such disengagement as being cancer-like in its disabling effect, one not simply weakening us, but embodying death itself at the very center of ourselves.

We can know theologically that we are eternally condemned to death, a death that becomes physically actual in cancer, and if cancer is a uniquely human disease, or so we imagine, it is here that we can actually taste our mortality, and thereby taste our eternal condition. Only Christianity knows that condition through original sin, an original sin that is an original fall, a fall only known in its totality by Christianity, just as Nietzsche could enact Christianity as the only absolute No-saying in the world. Surely cancer can stand forth as a genuine symbol of Christianity, one embodying an eternal death only interiorly known in Christianity, a death that is our deepest condition, or our deepest fallen condition.

The very actuality of cancer is the actuality of death itself, but a death only known in Christianity, or only celebrated in Christianity, as enacted in the Eucharist itself. Yet the Christian Eucharist is the eating of God, or the eating of the Son of God, and just as the Eucharist is the primal action or liturgy of Christianity, Christianity can be known as cancer-like in that very action, or in its uniquely Christian action. Is it cancer then that most concretely opens us to the very center of Christianity? Nietzsche would certainly accept this, and so, too, might Blake, and if a deeper Christianity is a profoundly kenotic Christianity, cancer could be understood as the most concrete way into Christianity, or into a deeper and kenotic Christianity.

Tom

Letter 71

To Friends (March 2, 2015)

Friends,

I wish to share with you a fundamental shift that I am undergoing with my manuscript on the genealogy of Nihilism. For I have been led by my own thinking to focus upon the liberation or the redemption effected by Nihilism as even embodied in a holy Nihilism. This is most openly or clearly present in Mahayana Buddhism as manifest in its ultimate identification of nirvana and samsara—even Nietzsche was excited by this—and it is clearly revolutionary. But our imaginative traditions embody nihilistic figures who are liberating powers as in Mephistopheles, Kirilov, characters in the dramas of Chekhov and Beckett, and even in Shakespeare, such as the fool in *King Lear*.

Hegel's most nihilistic writing occurs in his brief treatment of Comedy in the *Phenomenology*, and yet it is here that he speaks most forcefully of redemption, a redemption that is a feast of the gods. And it is Nietzsche, our foremost thinker of Nihilism, who in responding to Nihilism both in *Zarathustra* and *The Antichrist* can write most deeply about redemption. Is it Nihilism itself that most opens us to the depths of liberation? And does this occur again and again in late modernity? Recalling the symbol of an absolute grace that can occur only in the ultimate depths of sin and darkness, as manifest in the Christian symbol of the Descent into Hell, we can also recall that this is a rare and marginal symbol, and one that is seemingly never enacted in a major work of the imagination.

The Buddhist symbol of the identification of nirvana and samsara parallels the radical Christian symbol of the identification of Heaven and Hell, and if this is clearest in Blake, it is Blake who has given us the most comprehensive enactment of redemption itself. Moreover, this can be understood as a nihilistic enactment, for it empties or reverses every given symbol and ground so as to effect the explosion of an absolute apocalypse. This is an explosion that can be understood as the renewal of an original Christianity, an apocalyptic Christianity embodying that

Kingdom of God that is both the ending and the total transfiguration of history and the world.

Jesus is unique in proclaiming that Kingdom, a proclamation that is both proclamation and enactment, even if both are wholly transformed by ancient Christianity. Radical Christians such as Blake understand that transformation as a total inversion of genuine Christianity, and while even today nothing is more hidden and disguised than radical Christianity, in the perspective of radical Christianity an established Christianity and the established world itself are truly nihilistic. Even the voice of Nietzsche can be understood as a radical Christian voice, for it is the established Christian God who is the deification of nothingness and the will to nothingness pronounced holy!

This is just what made possible the late Blake's identification of the Creator and Satan, one occurring first in *Milton*, where Satan is unveiled as absolute sovereignty, and that sovereignty addressing us in the very voice of the Creator. Hence on the eleventh plate of the illustrations to the Book of Job[1] the Creator is portrayed with a cloven foot. True the ancient Gnostics, or the more radical among them, identified the Creator as absolute Evil, but Blake absolutely reverses the Gnostics in his exaltation of Body and Energy, an energy that is the absolute inversion and reversal of all bodiless Spirit. So it is that it is bodiless Spirit that is the true body of nihilism, a Spirit that is truly universal in late modernity, and a spirit that is the spirit of all of our established worlds.

Tom

Letter 72

To Friends (March 10, 2015)

Friends,

I am tempted to return to an old problem that I suspect may have returned, and that is the relation between God and Being in Aquinas, a problem that I suspect has never been resolved. This evoked an enormous controversy in the sixteenth century, perhaps taking us to the very center of the Catholic–Protestant controversy, and certainly having enormous pragmatic dimensions as philosophical problems seldom do. Is this a controversial question today? I sense that few Catholics are any longer interested in the question of God, and the question of Being and God is so utterly abstract as seemingly being without a human dimension.

Yet has not Heidegger elevated the question of Being to enormous heights, and apparently done so while ignoring Aquinas, even though he himself was originally a Thomist? Seemingly Being is a wholly abstract category, and with no "existential" bearing whatsoever, but we know better than that, as Sartre so deeply demonstrated. Indeed, Sartre's integral relation of Being and Nothing itself had deep effects, effects unimaginable today, which does say something about the dullness or hollowness of our world. Now I realize that a deep moratorium has long been established on the very word *God*, and one that I have come to believe is invisibly powerful even in Evangelical circles, and in my debates with evangelicals they were most defensive at this very point.

Many breathed a sigh of relief at the death of God movement with the hope that now God would disappear, and in a certain sense that did occur. My amateurish search for literature on God and Being did turn up little, but one astute book does seriously discuss it even while conceding that there is little substantial literature here. Sometimes I think that real Thomists are now silent and invisible, certainly far more silent than they once were, and yet the modern Catholic world until only recently was wholly Thomist, and Thomas himself was virtually deified. Have you encountered transcendental Thomism?—which I take seriously—

yet I think of it as a nontheological Thomism, a Thomism that in its absorption of modern philosophy bracketed the question of God.

Could it be that Catholics are now withdrawn from the question of God and Being? Is it possible for anyone today to be engaged in the problem of God and Being? Or despite Heidegger actually engaged with the question of Being? I have come to believe that the Heideggerian question, Why is there any Being at all, why not far rather nothing, is a deep reversal of Leibnitz and at bottom a question evoking Nihilism. But if so, does not this primal question simultaneously evoke both God and Being? And if that is true then we continue to be open to the question of Being and God, and perhaps our very Nihilism inevitably poses this question, and does so if only to make Being possible.

In Peace,
Tom

Letter 73

To Friends (April 11, 2015)

Friends,

The question of the radical Tillich is a deeply challenging question to theology itself for it questions the very possibility of a radical theology. That the young and Religious Socialist Tillich was a radical we can accept, but is such radicalism possible after this historical era? And if it can be demonstrated that Tillich abandoned his political radicalism after the demise of Marxism, is it possible that radical theology itself can be preserved after that demise? If it also can be demonstrated that Tillich finally constructed a phenomenological epoche upon the question of God, an epoche removing the question of God from theology itself, is this an inevitable consequence of any radical theology?

Nothing is more challenging in Tillich's theology than the question of God. The famous Taubes critique of Tillich[1] posed this question decisively, and in doing so raised the question of the very possibility of God in a genuinely modern theology. Tillich was never happy with this question, and despite innumerable efforts I was never able genuinely to discuss it with him, it is as though he was immune to this question, and immune in his deepest ground. True, Tillich always insisted to me that the real Tillich is the radical Tillich,[2] and he refused to place any distance between my theological atheism and his own theology, just as in public he refused to distance himself from me.

After his death I spent much time with his wife, Hannah, visiting her regularly in East Hampton, and joining her social world. And Hannah was genuinely radical in ways that Paulus was not, and most clearly so in the imaginative realm, here Paulus like every major theologian was closed to a radical imagination, which can be demonstrated in a universal theological closure to Blake. In a great many ways I reached an understanding of the uniqueness of my own theology by contrasting it with Tillich's, and most so just in this arena, and if I became a radical theologian by becoming a Blakean theologian, Blake is our supreme apocalyptic visionary, just as it was Blake who most deeply discovered the apocalyptic Jesus.

One of the high points in my life was participating in the famous Eliade–Tillich seminar on the History of Religions and Systematic Theology,[3] and afterwards the three of us would engage in deep theological discussions, in which Tillich was unusually candid, just as was Eliade himself. Eliade was far more deeply theological than he would commonly reveal, and in that context Tillich himself became more deeply theological, and I often wished that this could have been recorded and published. For the question of God was actually a deep one for Tillich, even if this is disguised in his writing, but it does come forth in his earlier writing, as so brilliantly edited by James Luther Adams.[4]

Many of us closest to him regretted the publication of Tillich's *Systematic Theology*, and particularly so its concluding volumes, which so disguise his genuine radicalism, and this even led to the question as to whether any systematic theology can be a radical theology. At no point is Catholic theology more disappointing than here, and thus the question arises as to whether any form of Christianity can be genuinely radical. I think that it is Tillich who poses this question most acutely, and it was Tillich who had the deepest impact upon radical theologians, and even if Tillich could not call together a Tillichian school of theology, he was an ultimate theological presence, and above all so in so forcefully posing the very question of a radical theology.

Thomas J. J. Altizer

Letter 74

To Friends (April 26, 2015)

Friends,

In reading an excellent essay by Cyril O'Regan on the Augustinianism of Benedict XVI,[1] I am driven to look again at that Augustinianism that so deeply shaped my own theology.[2] For it is a universe removed from all conservative Augustinianism as so deeply embodied in Benedict, and one ground of this difference is the deeply historical ground of my Augustinianism, as opposed to the ahistorical approach of Benedict. Benedict is closed to the profound historical transformations effected by Augustine, such as the Reformation, and before that the creation of Western Christianity itself, which was most decisively effected by the impact of Augustine. The West is so ignorant of Eastern Christianity that it simply ignores the differences between Eastern and Western Christianity, but these are profound and of ultimate importance.

One of these is the mystical or the purely mystical itself, already Augustine deeply transforms the mystical, and does so as the first major Western mystical theologian, but one who integrates the mystical and the non-mystical, as apparently never occurs in the East. So, too, Augustine is apocalyptic and non-apocalyptic at once, a point that O'Regan stresses, and while this is paralleled in the East, there it is never so decisive as in Augustine. Indeed, there is a theological universality in Augustine that is unique, as can concretely be demonstrated in the large number of Augustinian movements and theologies, as opposed to what is apparently a near monolithic Thomism. While it is true that a new Thomism occurs in a modern transcendental Thomism, that Thomism in absorbing modern German philosophy never deviates theologically from a traditional Thomism, whereas Augustinian theologies as such realize very different theological expressions.

Clearly Augustinianism is far more ecumenical than Thomism, as witness its profoundly Protestant expressions, and even if Thomism has occurred in conservative Anglicanism and conservative Lutheranism, it has never occurred in any modern expressions of Protestantism. If modernity is born in Luther, as Hegel believes, this is clearly a rebirth of Augustine, and a rebirth of a radical Augustine, that occurs not only in Luther and Calvin, but in Descartes, too, a Descartes who can even be understood as an Augustinian theologian, although such a rebirth and renewal is far more profound in Hegel. Catholic theologians long battled modern philosophy as an expression of Protestantism, but it is an expressions of Augustinianism, too, and particularly so insofar as it revolves about a radically new subject, but a subject that is a rebirth of the Augustinian subject.

O'Regan demonstrates that Benedict simply ignores all radical interpretations of Augustine, but these occur again and again in Augustinian scholarship, which is yet another chasm between Augustinianism and Thomism. Perhaps the primary center of an Augustinian hermeneutics is subject itself, a subject that was actually discovered by Augustine, for it never occurs or is realized in the Classical world. This is an event of extraordinary significance, for it makes possible a uniquely Western consciousness and world, and while it is not until Hegel that subject becomes absolute philosophically, it is already dominant in a uniquely Western art and literature. Now if we can apprehend a decisive role of Augustine or Augustinianism in the post-Classical Western imagination, we can apprehend a truly absolute importance of Augustine, but one wholly closed to ecclesiastical authority.

While Augustine was a great church theologian, and even could be known as creating a deeper church theology, it is just here that modern Augustinianism is most ambivalent, establishing yet another chasm between Augustinianism and Thomism. This ambivalence can be observed in the large number of non-church Augustinians, a body including the most creative Augustinians, for even if Hegel was a practicing Lutheran, the church and church tradition are wholly missing from his philosophy, and major Hegel scholars such as O'Regan have clearly demonstrated his total heresy.

Not only is such a heresy impossible in Thomism, it is virtually universal in modernity itself, leading to profound Catholic assaults upon modernity, as occurs in every modern pope. Only a resolution of this controversy can bring peace to Catholicism, as attempted in Vatican II, but Augustinianism could effect such a peace, for it is ultimately powerful both in Catholicism and modernity, and virtually alone as such.

Nor can we neglect as Benedict does Augustine the writer, the only great writer theologian, who apparently produced more written work than any other theologian, and the only theologian to have written a literary classic, the *Confessions*. Despite claims to the contrary, Augustine created the literary genre of autobiography, for there is simply nothing comparable to his until centuries later, and that very creation speaks volumes about his discovery of a new interiority. It could be said that Augustine created psychology itself, for Classical psychologies are external rather than internal psychologies, for the interior itself was unknown and unmanifest until the advent of Christianity. This is an ultimate revolution that is almost universally ignored, and it must be ascribed to Paul, who was the first to give witness to the depths of our interior, and at this absolutely crucial point Augustine is our first Pauline theologian.

One of the decisive signs of the absolute genius of Paul is his overwhelming and continual impact, and this is true of Augustine, too, even if it is ignored by Benedict and the church hierarchy itself. Aquinas himself was an Augustinian, and a deep Augustinian, but the Hierarchy has been able to create a non-Augustinian Thomism, and one if only thereby a Scholastic Thomism. True, the Scholastics thought of themselves as Augustinians, but theirs was the most sterile of Augustinianism apart from the rebellions that occurred within it, yet these rebellions were the seeds of a future creativity, eventually leading both to Protestantism and to modern science.

It is odd to think of Augustine as a father of science, but he certainly was the father of the science of psychology, just as he was the father of that Cartesianism that became a scientific philosophy in phenomenology, and nothing comparable to this occurred to Thomistic thinking, or for that matter to any other theological thinking. Despite Benedict, it is simply impossible to be both an Augustinian and an isolated thinker, unless it is possible for the purely ecclesiastical thinker. Benedict as a former grand inquisitor does have a special route into Augustine, for no major theologian has been such an effective destroyer of heterodoxy as was Augustine, and certainly not Benedict himself.

Perhaps Augustine was most important as a destroyer of heterodoxy in his destruction of the Pelagians, Pelagians who had absolutely rebelled against Paul, and thereby had rebelled against an absolute grace, The Pelagians can even be understood as common sense Christians, insisting upon the actuality of freedom and virtue, and of the dignity of human life. Augustine thought of them as absolute destroyers of Christianity in their refusal of the absolute necessity of grace, a refusal that is finally a refusal of redemption in Christ. Hence Bishop Augustine destroyed them, but he destroyed them as a Bishop and not as jailer, thereby serving what he knew and obeyed as the Catholic Church.

Yet this was the time when the Catholic Church was facing its first absolute challenge, the challenge of the Barbarian invasion of the Roman Empire, an invasion soon occupying North Africa, and there concretely challenging Bishop Augustine. All of Augustine's pastoral skills were now called upon and somehow he met this challenge, a challenge never so totally occurring in our world. And despite the Dark Ages that soon followed, the ancient church not only survived but evolved, and even if this did not decisively occur unto the twelfth century, occur it did, and very much under the impact of Augustine. No other Church Father has had such an ultimate effect, and just as Augustine can justly be known as the Father of Western Christianity, he can thereby be known as an ultimate Father of the West itself.

Tom

Notes

Introduction

1. Thomas J. J. Altizer, *The Gospel of Christian Atheism* (Philadelphia: Westminster, 1966), p. 22.
2. Mark C. Taylor, "Foreword: The Last Theologian," in Thomas J. J. Altizer, *Living the Death of God: A Theological Memoir* (Albany: State University of New York Press, 2006), p. xi.
3. Altizer, *Living the Death of God,* p. 177. Hereafter this book is cited as LDG.
4. Felix Pryor, ed., *The Faber Book of Letters* (1988; reprint, London: Faber & Faber, 1990), p. x.
5. Brian Schroeder, e-mail to Mike Grimshaw, June 22, 2015.

1 To Brian Schroeder (October/November 1996)

* Brian Schroeder, Professor of Philosophy at Rochester Institute of Technology. At this time Schroeder was in the Department of Philosophy, University of Southern Maine.

1. Brian Schroeder, *Altared Ground: Levinas, History, and Violence* (New York: Routledge, 1996).
2. The chapters are entitled "BackGround," "UnderGround," "HelioGround," "IdeoGround," "MiddleGround," "WarGround," "MysticGround," "TransGround," "DiaGround," and "HyperGround."

2 To Brian Schroeder (December 5, 1996)

1. Eric Voegelin (1901–1985), German-born American political philosopher.
2. Leo Strauss (1899–1973), Jewish German-American political philosopher. His thought and work spanned ancient Greece to Nietzsche. Primarily a historian of

philosophy, he focused on premodern thought and applied it to contemporary issues. A central issue of his thought is his use of the term "theologico-political predicament" to describe what he saw as the ongoing and devastating consequences of the early modern attempt to separate theology from politics. This arose from his work on Spinoza. See: https://leostrausscenter.uchicago.edu/

3. Robert P. Scharlemann (1929–2013), philosophical theologian and Tillich scholar who taught at the University of Iowa and University of Virginia. For Altizer's appreciation of Scharlemann, see Thomas J. J. Altizer, *Living the Death of God* (Albany: State University of New York Press, 2006), p. 23.
4. Paul's terms for *flesh* and *spirit* in the New Testament epistles.
5. To sublate: an elevation that contains within it both the thesis and antithesis that survive as moments. Hegel tends to use the term in three senses at once: to raise; to abolish or cancel; and to keep or preserve. Hegel applies the term to both things and concepts in the sense that what is earlier or prior is both destroyed and preserved in what is sublated.

3 To D. G. Leahy (January 17, 1997)

* D. G. Leahy (1937–2014), philosophical theologian, professor of classics and religion at New York University, distinguished visiting professor at Loyola University of Maryland, and founder of New York Philosophy Corporation. This letter discusses ideas in Leahy's then recently published *Foundation: Matter the Body Itself* (Albany, NY: State University of New York Press, 1996). For Altizer's appreciation and assessment of Leahy, see Altizer's *Living the Death of God: A Theological Memoir* (Albany, NY: State University of New York Press, 2006), pp. 51–52.

1. The coincidence of opposites, an idea can be traced to the pre-Socratic philosophers. Its use as a contemporary religious-studies term arises from Nicholas of Cusa (1401–1464) and his essay *De docta ignorantia* (1440), and was taken up by scholars of religion and mysticism in the twentieth century.
2. French term for resentment. The use of this in philosophy and theology arises from Nietzsche in *On the Genealogy of Morals*. It refers to both resentment and hostility expressed as a form of revenge. *Ressentiment* is an experience that has become repressed and internalized and yet generates the expression of public values.
3. St. John of Damascus (c. 676–754?/787). Challenged the iconoclastic policy of Byzantine Emperor Leo III arguing for the veneration but not the worship of icons; that the icon enables devotion to reach the unseen recipient, and that subsequent to the incarnation the icon is a profession of faith.
4. The "thinking now occurring for the first time" is one of Leahy's key philosophical ideas; see his book *Foundation*, pp. 5–6, and references in the index.
5. The fall of the Berlin Wall and the beginning of the end of the Eastern bloc and Soviet Empire.

4 To Lissa McCullough (June 13, 1997)

* Lissa McCullough, co-editor with Brian Schroeder of *Thinking through the Death of God: A Critical Companion to Altizer* (Albany: State University of New York Press, 2004), and author of *The Religious Philosophy of Simone Weil* (London: IB Tauris, 2014). At this time she was assistant professor of religious studies at Muhlenberg College.

1. Herman Melville (1819–1891); American novelist and poet best known for his novel *Moby-Dick* (1851). Melville and *Moby-Dick* play an important role in Altizer's assessment of American religion and culture, and questions of God in modernity more generally. Altizer taught in the English department at SUNY Stony Brook from 1968–1995, where he established an interdisciplinary program in Religious Studies.
2. Franz Kafka (1883–1924), Prague-born German-language Jewish writer. The world of Kafka's writing is one of alienation, struggles with illogical bureaucracy, transformations, damaged relationships, and the existential questions of modern existence.
3. This is a collection of writings and illustrations dealing with humanity's hope for immortality. Ranging from Plato and the Bible to the twentieth century, it also includes 57 illustrations from master works of art. Grierson (1866–1960) was a Scottish academic, literary scholar and critic.
4. Literally "bad tidings" from Nietzsche's *The Antichrist* (1895). In this, Nietzsche infamously states there was only one Christian and he died upon the cross. The post-crucifixion Christ of what becomes Christianity is named as *dysangel*, the reverse of how the evangel had lived.
5. William Blake,(1757–1827), English poet, printmaker and painter and a central prophetic theological influence upon Altizer.
6. See Eric Auerbach, *Mimesis: The Representation of Reality in Western Literature* (Princeton: Princeton University Press, 1953).
7. Samuel Beckett (1906–1989), Irish anti-realist novelist, playwright and poet who lived and wrote mostly in Paris. His works offer a bleak, modernist, tragicomic view of what can be termed the absurdity of human existence.
8. Meister Eckhart (1260–1328), German Dominican priest, mystic, theologian and philosopher.
9. Simone Weil (1909–1943); French philosopher and activist, born into a Jewish agnostic family she moved toward an unbaptized Christian mysticism from the mid-1930s.

5 To Brian Schroeder (January 18, 1998)

1. This refers to "the turn" in Heidegger's philosophy from 1930 to early 1940s away from *Being and Time* (1927) toward the question of "the essence of truth" and the way in which language can unfold the question of being. What it actually meant for Heidegger is the subject of much ongoing debate.

2. Reiner Schürmann, *Heidegger on Being and Acting: From Principles to Anarchy* (Bloomington: Indiana University Press, 1986).
3. Edward S. Casey, *The Fate of Place: A Philosophical History* (Berkeley: University of California Press, 1997).
4. Paul Ricoeur, *The Symbolism of Evil* (New York: Harper & Row, 1967).
5. Emmanuel Levinas, "Transcendence and Evil," in *Collected Philosophical Papers*, trans. Alphonso Lingis (Dordrecht: Marinus Nijhoff, 1987), pp. 175–186; original French version, "Transcendence et Mal," *Le Nouveau Commerce* 41 (1978): 55–75.
6. Ray Monk, *Wittgenstein: The Duty of Genius* (London: Jonathan Cape, 1990).
7. Husserl to Professor Paul Natorp of Marburg University, October 8, 1917.
8. Jacques Derrida, *The Gift of Death* (Chicago: University of Chicago Press, 1996).
9. Edward S. Casey, Distinguished Professor of Philosophy at State University of New York, Stony Brook. See: http://edwardscasey.com/
10. Cornelio Fabro, *God in Exile: Modern Atheism; A Study of the Internal Dynamics of Modern Atheism, From Its Roots in the Cartesian Cogito to the Present* (Westminster, MD: Newman Press, 1968); original edition, *Introduction to Modern Atheism* (Rome: Studium, 1964). For an overview of Fabro's life and work see: http://www.corneliofabro.org/en/documento.asp?ID=24

6 To Brian Schroeder (January 28, 1998)

1. Philippe Nemo, *Job and the Excess of Evil* (1978; Pittsburgh: Duquesne University Press, 1998).
2. The quotation is from "Transcendence and Evil," in Emmanuel Levinas, *Collected Philosophical Papers*, trans. Alphonso Lingis (Dordrecht: Marinus Nijhoff, 1987), p. 179.
3. That is, the "happy fault." *Felix culpa* in theology is a way to describe the redemptive necessity of the Incarnation and therefore a way to describe the Fall. It is also a way to engage with issues of theodicy in that a bad or evil event or action may result in an unimagined good consequence or outcome.
4. Levinas, "Humanism and Anarchy," in *Collected Philosophical Papers*, pp. 127–139. The following discussion refers to this essay.
5. According to Hegel's *Science of Logic* this is the endless series of finite items wherein a finite 'something' requires a concept of a finite 'something else' that in turn becomes a further finite 'something' requiring a further finite 'something else,' on to infinity. This creates a 'bad infinity' progression of finite limitations.

7 To Ray L. Hart (April 13, 1998)

*Ray L. Hart, Professor of Religion and Theology, Boston University School of Theology.

1. This letter responds to an early draft of Ray L. Hart, *God Being Nothing* (Chicago: University of Chicago Press, forthcoming in 2016).

2. Jacob Boehme (1575–1624), German Lutheran mystic and theologian.
3. Robert Cummings Neville, Professor of Philosophy, Religion, and Theology, Boston University School of Theology.
4. Robert C. Neville, *God the Creator: On the Transcendence and Presence of God* (Chicago: University of Chicago Press, 1968; reprinted with corrections and a new preface, Albany: State University of New York Press, 1992).
5. http://www.pbs.org/wgbh/pages/frontline/shows/religion/
6. Jack Miles, *God: A Biography* (New York: Knopf, 1995), was awarded the Pulitzer Prize for biography in 1996. Miles, a former Jesuit seminarian, is a journalist, writer, and academic.
7. Creation from the very beginning or source of origin.
8. Continuous and ongoing creation by the Creator.
9. John Milbank, Professor in Religion, Politics and Ethics, University of Nottingham. At this date Milbank was a fellow at Peterhouse, Cambridge and was soon to move to University of Virginia to become Frances Myers Ball Professor of Philosophical Theology (1999–2004).
10. Radical Orthodoxy, a theological movement established by Milbank, Graham Ward, and Catherine Pickstock arising from the combination of using postmodernism and St. Augustine to critique modernity. Its central argument is a reclaiming of theology as the queen of the sciences as the central necessary starting point to understand the world. It took an early expression in John Milbank, Catherine Pickstock, and Graham Ward, eds., *Radical Orthodoxy: A New Theology* (New York: Routledge, 1998).
11. Friends of God: a German medieval mystical group of laity arising out of the teaching of Meister Eckhart.
12. Thomas J. J.Altizer, *The Genesis of God: A Theological Genealogy* (Louisville, Kentucky: Westminster John Knox Press, 1993).
13. Uproar or disturbance.
14. Disturbing energy.
15. *Trieb* (urge): the primordial struggle of the primal urge that, as inner urge, seeks an outer satisfaction, realization, or identity.
16. The Kabbalistic school derived from Rabbi Isaac Luria (1534–1572).
17. Valentinus (100–c. 160/170?), Egyptian Christian mystic and gnostic theologian.
18. Gnostic text written by school of Valentinus and discovered at Nag Hammadi in 1945. See: http://gnosis.org/naghamm/got.html.
19. Eric Voegelin (1901–1985), German-born American political philosopher. Voegelin is referenced here for his works that criticized what he took to be a gnostic influence in politics. His "metaxy," taken from the "middle ground" in Plato's *Symposia,* is redefined by Voegelin as where humanity permanently finds itself between two poles of existence.

8 To Thomas A. Carlson (December 30, 1998)

*Thomas A. Carlson, Professor of Religious Studies, University of California, Santa Barbara.

1. Thomas A. Carlson, *Indiscretion: Finitude and the Naming of God* (Chicago: University of Chicago Press, 1999).

2. Jean-Luc Marion, philosopher and Catholic theologian.
3. Hans Urs von Balthasar (1905–1988) Swiss Catholic priest and theologian.
4. Pseudo-Dionysius the Areopagite (c. fifth–c. sixth century), Christian Neoplatonic philosopher. and theologian, called Pseudo-Dionysius for the false attribution of being the convert Dionysius of Acts 17:34.
5. The limitless.
6. Anaximander (c. 610–546 BCE), pre-Socratic Greek philosopher. He is taken to provide the earliest use of *apeiron* to apply to the original principle, which he designated as the boundless or unlimited origin of all things.
7. In Hegel, *aufgehoben* is the sublation when the thesis and antithesis interact into that which transcends what was and was not.
8. Ten elegies, written 1912–1922, by Rainer Maria Rilke (1875–1926). These existential poems, lamenting modern existence, are intensely mystical and religious in tone, making great use of the symbolism of angels who are utterly indifferent to mankind.
9. Orrin Summerell, "The Othernes of the Thinking of Being: Heidegger's Conception of the Theological Difference," in *The Otherness of God*, ed. Orrin F. Summerell (Charlottesville: University of Virginia Press, 1998), pp. 111–134.

9 To D. G. Leahy (April 4, 2000)

1. D. G. Leahy's lecture at Loyola University at Baltimore as distinguished visiting professor, March 2000. The lecture is published as the appendix to Leahy's book *Faith and Philosophy* (New York: Ashgate, 2003) entitled "Thinking in the Third Millennium: Looking Without the Looking Glass," pp. 143–163.
2. In his youth Leahy authored an epic poem, which he later decided to destroy.
3. Charles Sanders Pierce (1839–1914), American pragmatist philosopher, scientist, and logician.

10 To Cyril O'Regan (April 12, 2000)

* Cyril O'Regan, Huisking Professor of Theology, University of Notre Dame.

1. Nicholas Berdyaev (1874–1948), Russian religious and political philosopher. Upon his exile to Berlin, then Paris in 1922, his version of Christian existentialism was widely influential in the mid-twentieth century, especially his work on freedom and the possibility of a new Middle Ages.
2. All three form Dante's epic poem *Divine Comedy* (c. 1308–1320). In this allegoric poem Dante (c. 1265–1321) recounts an imaginative journey toward God via Hell, Purgatory, and then Paradise.
3. *Paradise Lost* (1667), an epic poem by John Milton (1608–1674) concerning the Fall of Man.
4. *The Divine Comedy*.
5. James Joyce, *Finnegans Wake* (London: Faber & Faber,1939).

6. Hans Urs von Balthasar, *Mysterium Paschale: The Mystery of Easter* (Grand Rapids: W. B. Errdmans, 1970; English translation 1990).
7. Gilles Deleuze, *Spinoza et le problème de l'expression* (Paris: Editions de Minuit, 1968); translated into English as *Expressionism in Philosophy: Spinoza,* trans. Martin Joughin (New York: Zone books, 1992); and *Spinoza* (1970), republished as *Spinoza: Philosophie pratique* (Paris: Editions de Minuit, 1981), translated into English as Deleuze, *Spinoza: Practical Philosophy,* trans. Robert Hurley (San Francisco: City Lights, 2001).
8. Deleuze, *Spinoza: Practical Philosophy* (1970).

11 To Gan Yang (September 8, 2001)

*Gan Yang (b. 1952), Chinese intellectual noted for the influential book series he edited in the 1980s in Beijing entitled *Culture: China and the World,* intended to catalyze a new critical-intellectual culture in China by the diffusion of modern Western social and intellectual theory, including works of Cassirer, Weber, Husserl, Heidegger, Buber, and scores of others.

1. An unpublished confessional document, shared with friends, in which Altizer discusses formative personal experiences, many of which are briefly alluded to in *Living the Death of God.*
2. Lissa McCullough became friends with Gan Yang at the University of Chicago in the early 1990s, where he went to study in the wake of the Tiananmen Square crackdown on pro-democracy protesters in 1989.
3. Martin Heidegger, *Beiträge zur Philosophie [Contributions to Philosophy (From Enowning)],* written 1936–1938; published in German 1989; English translation 1999. (Bloomington, IN.: Indiana University Press, 1999).

12 To Lissa McCullough (September 7, 2005)

1. Hurricane Katrina struck the Gulf coast of the United States on August 29, 2005. In New Orleans the levees breached and severe flooding occurred. The economic impact has been estimated at over $108 billion. At least 1,836 people died as a result of the Hurricane. The United States government was severely criticized for its response to the hurricane and its aftermath, especially in New Orleans.
2. Christopher Hill, *The World Turned Upside Down: Radical Ideas during the English Revolution* (London: Maurice Temple Smith, 1972; London: Penguin, 1991).
3. Gerrard Winstanley (1609–1676), English Protestant radical and founder of the Diggers. The Diggers, originally self-named the True Levelers, sought a society of communal ownership and true equality arising from their reading of the Bible. They were named the Diggers from their practice of farming common land. While

they were radical Protestants and deeply theological, Winstanley and the Diggers have also been claimed as proto-communists.

4. Thomas J. J. Altizer, *History as Apocalypse* (Albany, NY: State University of New York Press, 1985).
5. Thomas J. J. Altizer, *Total Presence: The Language of Jesus and the Language of Today* (New York: Seabury, 1980).
6. Milton's *On Christian Doctrine* is a theological treatise that was lost after his death in 1674 and was only rediscovered in 1823.
7. President George W. Bush, forty-third President of the United States of America (2001–2009).
8. In his memoir *Living the Death of God*, Altizer writes: "Once the death of God controversy fully broke, I was continuously on television. As I traveled about the country I was invariably interviewed for local television, and I always responded with preaching; perhaps I was the first television evangelist, and I preached by declaring that everything we know as God is dead, and that this death is the gospel, is the 'good news', I would pronounce this with as much passion as possible, and proclaim it as the full and final advent of an absolute joy" (p. 16).

13 To Lissa McCullough (October 13, 2005)

1. Ingmar Bergman (1918–2007), Swedish screenwriter and filmmaker.
2. William Blake (1757–1827), English poet, printmaker, and painter and a central prophetic theological influence on Altizer.
3. Mark C. Taylor, Professor, Department of Religion, Columbia University, who at that time was professor at Williams College.
4. Thomas J. J. Altizer, *Living the Death of God*. (Albany, NY: State University of New York Press, 2004).
5. In his 2002 State of the Union address, President George W. Bush used the term "axis of evil" in reference to Iran, Iraq, and North Korea. Accused of seeking weapons of mass destruction and supporting terrorism, the term axis of evil was used to underpin what became known as the "war on terror." The phrase was coined by Bush speechwriter David Frum.

14 To Edith Wyschogrod (December 2, 2005)

* Edith Wyschogrod (1930–2009), American philosopher of religion, J. Newton Rayzor Professor of Philosophy and Religious Thought at Rice University, where she taught 1992–2003. Her work focused on continental thought and ethics, and she played a central role in introducing the work of Levinas into American religious and philosophical thought.

1. Wyschogrod critiqued Levinas's infinite as a Hegelian "bad infinite." See Edith Wyschogrod, *Emmanuel Levinas: The Problem of Ethical Metaphysics* (1974; reprint, New York: Fordham University Press, 2000), p. 239. Derrida's deconstruction was also attacked as an expression of the bad infinite.
2. Edith Wyschogrod, "Incursions of Evil: The Double Bind of Alterity," in *Modernity and the Problem of Evil,* ed. Alan D. Schrift (Bloomington: University of Indiana Press, 2005), pp. 188–198.
3. Gregory Bateson's double bind theory (1956) attributed a predisposition to schizophrenia to the confusion created by contradictory familial interaction as a form of control. Bateson (1904–1980) was a polymath anthropologist, scientist, and philosopher.
4. Dominique Janicaud (1937–2002), French philosopher whose work focused on critical engagement with Heidegger.
5. Jean Baudrillard (1929–2007), French trans-disciplinary theorist whose work covered philosophy, sociology, cultural theory and politics. Perhaps best known for tracing and debating the impact of technology on modern and postmodern life and thought.
6. Edith Wyschogrod, "The Warring Logics of Genocide," in *Crossover Queries: Dwelling with Negatives, Embodying Philosophy's Others*, ed. Edith Wyschogrod (New York: Fordham University Press, 2006), pp. 222–235.
7. *Il y a:* the "there is," Levinas's term for the impersonal background of existence.

15 To D. G. Leahy (December 29, 2005)

1. Alain Badiou, *St. Paul: The Foundation of Universalism*, trans. Ray Brassier (1997; Stanford: Stanford University Press, 2003).
2. Badiou is actually a Maoist.
3. Alain Badiou, *Infinite Thought: Truth and the Return to Philosophy*, trans. and ed. Oliver Feltham and Justin Clemens (London: Continuum, 2003).

16 To Brian Schroeder (January 23, 2006)

* Brian Schroeder, "Theological Nihilism and Italian Philosophy," *Philosophy Today* 49, 4–5 (Winter 2005): 355–361.

1. Franca D'Agostini, "Nihilism in Italy," *Philosophy Today* 49, no. 4 (2005): 342–354.
2. Sergio Givone, Professor of Aesthetics, Department of Humanities and Philosophy, University of Florence.
3. Brian Schroeder and Silvia Benso co-edit the State University of New York Press series in Contemporary Italian Philosophy. Givone contributed "Philosophy, Poetry and Dreaming," to Benso and Schroeder, eds., *Contemporary Italian Philosophy: Crossing the Borders of Ethics, Politics, and Religion* (Albany, NY: State University of New York Press, 2007), pp. 33–46.

17 To Walter A. Strauss (February 6, 2006)

*Walter A. Strauss (1923–2008) scholar of comparative literature, formerly Treuhaft Professor of the Humanities, Case Western Reserve University, and earlier a colleague of Altizer's at Emory University.

1. Herbert Weisinger, (1913–1999), professor of English at Michigan State University, author of *Tragedy and the Paradox of the Fortunate Fall* (Lansing: Michigan State College Press, 1953).
2. French silent movie (1928) acknowledged as a cinematic classic. Dreyer (1889–1968) was a Danish film director. His films often focus on religious themes and at the time of his death he was working on a film about Jesus.

18 To Lissa McCullough (February 7, 2006)

1. Thomas J. J. Altizer, *Godhead and the Nothing* (Albany, NY: State University of New York Press, 2003). See chapter 2.
2. Adolf von Harnack, *Monasticism: Its Ideals and History and the Confessions of St. Augustine* (1901).
3. Erich Auerbach, *Dante: Poet of the Secular World* (Chicago: University of Chicago Press, 1961; reprint, NYRB Classics, 2007).
4. Erich Auerbach, *Mimesis: The Representation of Reality in Western Literature* (Princeton, NJ: Princeton University Press, 1953).

19 To Friends (March 2, 2006)

1. Joseph Ratzinger, Pope Benedict XVI from 2005 until his resignation in 2013, author of *Introduction to Christianity* (London: Burns & Oats, 1968); revised English edition with new introduction, Ignatius Press, 2000.
2. Karl Adam (1876–1966), German Catholic theologian whose *Spirit of Catholicism* (1924) is an introduction to the Catholic Church.
3. Pierre Teilhard de Chardin S. J. (1881–1955), French Jesuit Priest, philosopher and paleontologist. His writings are a combination of mysticism, cosmology and evolutionary thought.
4. Adolf von Harnack, *History of Dogma* (7 vols. in English translation, 1894–1899).

20 To Lissa McCullough (March 9, 2006)

1. Thomas J. J. Altizer, *Genesis and Apocalypse: A Theological Voyage toward Authentic Christianity* (Louisville: Westminster John Knox, 1990).

2. Bernard McGinn, Naomi Shenstone Donnelley Professor Emeritus of Historical Theology and the History of Christianity, University of Chicago, Divinity School.
3. Adolf von Harnack, *Marcion and the Gospel of the Alien God* (Leipzig: J. C. Hinrichs'sche Buchhandlung, 1920).
4. John Scotus Eriugena (c. 800–c.877), Irish-born Neoplatonist philosopher and translator.
5. Nagarjuna (c 150–250 CE), Indian Buddhist philosopher, noted as the most important after the Buddha himself. Central to his philosophy is the concept of emptiness (*śūnyatā*), the emptiness of a something that can be understood in terms of essence, substance, and absolute reality, as well as the emptiness of the way humans conceptualize a something.
6. Martin Heidegger, *Beiträge zur Philosophie [Contributions to Philosophy (From Enowning)], written 1936–1938;* published in German 1989; English translation 1999.
7. Thomas J. J. Altizer, *The Self-Embodiment of God* (New York: Harper & Row, 1977); reprint, with an introduction by Jacob Neusner (Lanham, MD: University Press of America, 1985).
8. Thomas J. J. Altizer, *History as Apocalypse* (Albany, NY: State University of New York Press, 1985).
9. University of Chicago Divinity School, 1951.
10. Martin Luther, *The Bondage of the Will* (1525).

21 To Friends (June 29, 2006)

1. Variously understood as self-surrender, releasement, resignation to God's will—or most simply a letting go; Heidegger borrowed this from Eckhart.

22 To Friends (July 5, 2006)

1. Rudolf Bultmann (1884–1976), German Lutheran New Testament scholar and theologian, best known for his existentially derived project of demythologization with respect to the New Testament.
2. Hidden being.

23 To Alina N. Feld and Lissa McCullough (September 19–22, 2006)

*Alina N. Feld, author of *Melancholy and the Otherness of God: A Study of the Hermeneutics of Depression* (Lanham, MD: Lexington, 2011). During her undergraduate years Feld studied with Altizer at SUNY at Stony Brook.

1. Thomas J. J. Altizer, *Godhead and the Nothing* (Albany, NY: State University of New York Press, 2003).
2. Stephane Mallarme (1842–1898), French symbolist poet concerned with tension between this material world and the world of ideal forms. He believed there was an infinity found in the void; that is the ideal world is concealed in the emptiness of space.
3. Wallace Stevens (1879–1955), American poet.
4. Written by Heidegger in 1936–1938, *Contributions to Philosophy* (*Beiträge zur Philosophie)* is taken to signal the turn in Heidegger's thinking from *Being and Time* (1927). Two English translations have been published since the first German edition appeared in 1989: *Contributions to Philosophy (From Enowning)*, trans. Parvas Emad and Kenneth Maly (Indiana University Press 1999); and *Contributions to Philosophy (Of the Event),* trans. Richard Rojcewicz and Daniela Vallega-Neu (Bloomington: Indiana University Press, 2012). The term *Ereignis* was rendered in the 1999 translation as the neologism "enowning," and in the 2012 translation as "the event."
5. The final play written by Sophocles in 406 BCE takes the middle position in his Theban plays between *Oedipus the King* and *Antigone.*

24 To John B. Cobb, Jr. (September 27, 2006)

*John Cobb, Jr. (b. 1925), American process theologian, philosopher, and environmental ethicist, cofounder of the Center for Process Studies in Claremont, California.

1. Monk, *Wittgenstein: The Duty of Genius,* mentioned in the letter of January 18, 1998, to Brian Schroeder.
2. Under the aspect or standpoint of eternity.

25 To John B. Cobb, Jr. (October 4, 2006)

1. Alfred North Whitehead, *Religion in the Making* (1926), four lectures delivered at Kings' Chapel, Boston, in which Whitehead applies his process thinking on the development of science to the development of religion.
2. This reference is to Whitehead's statement in Lecture 4, "Truth and Criticism": "The kingdom of heaven is not the isolation of good from evil. It is the overcoming of evil by good. This transmutation of evil into good enters into the actual world by reason of the inclusion of the nature of God, which includes the ideal vision of each actual evil so met with a novel consequent as to issue in the restoration of goodness." See: http://theology.co.kr/whitehead/religion/4.html
3. Thomas J. J. Altizer, *Godhead and the Nothing* (Albany, NY: State University of New York Press, 2003).
4. The reference is to Lecture 3, "Body and Spirit," specifically the section "God and the moral order": "A complete determinism would thus mean the complete self-consistency of the temporal world. This is the conclusion of all thinkers who are

inclined to trust to the adequacy of metaphysical concepts. The difficulty of this conclusion comes when we confront the theory with the facts of the world. If the theory of complete determinism, by reason of the necessity of conformation with the nature of God, holds true, then the evil in the world is in conformity with the nature of God." See: http://theology.co.kr/whitehead/religion/3.html.

26 To Andrew W. Hass (January 5, 2007)

*Andrew W. Hass, Reader in Religion, University of Stirling.

1. Giotto di Bondone (c. 1267–1337), Florentine painter and architect and a central forerunner to the Italian Renaissance.

27 To Andrew W. Hass (January 11, 2007)

1. Giorgio Agamben (b. 1942), Italian continental philosopher and radical political theorist.
2. Louis Althusser (1918–1990), French Marxist philosopher, highly influential in the 1960s and again more recently in the turn to post-Marxism.
3. A fire destroyed Altizer's home in Buck Hill Falls, Pennsylvania, in May 2004.
4. Conversations of Goethe with Johann Peter Eckermann (1836/1848).
5. See Giorgio Agamben, *The Man Without Content*, trans. Georgia Albert (Stanford University Press, 1999).
6. Darrel E. Christensen, ed., *Hegel and the Philosophy of Religion: The Wofford Symposium* (The Hague: Nijhoff, 1970). The symposium was held November 28–30, 1968, at Wofford College, South Carolina. Altizer responded to Murray Greene's paper "Hegel's 'Unhappy Consciousness' and Nietzsche's 'Slave Morality.'"

28 To Friends (January 24, 2007)

1. This paper, "The Unique Ground of America" was revised a number of times; the version included in this volume was presented in 2012 at a conference entitled "The Crisis of American Democracy," Fort Wayne, Indiana.
2. William Blake, *America: A Prophecy* (1793).

29 To Friends (January 25, 2007)

1. John Meyendorff, *Christ in Eastern Christian Thought* (1969; City: St. Vladimir's Seminary Press, 1975).

2. The doctrine that God can suffer.
3. Cyril of Alexandria (378–444 CE), Patriarch of Alexandria in the period 412–444 CE, led opposition to Nestorius, Patriarch of Constantinople, over the question of the nature of Christ.
4. *Theotokos,* which means "Bearer of God," was a term applied to Mary in opposition to Nestorius and his followers, who argued that there were two persons and two natures in Jesus, the divine and the human, and that Mary was only *Christotokos,* "Bearer of Christ."
5. The Greek word *phusis* means "nature" in the sense to grow or spring forth, or the inherent origin.
6. The underlying status or substance; in theology, the doctrine that Christ is one person with two natures, divine and human.
7. Gregory Nazianzen (330–390 CE), Archbishop of Constantinople, also known as Gregory the Theologian.
8. Gregory's full statement is: "We needed a God made flesh and put to death in order that we should live again."
9. Leontius of Jerusalem, a sixth-century theological writer and monk who wrote treatises against the Nestorians and the Monophysites.
10. Attributed to Leontius of Byzantium; there is debate as to whether he is the same individual as Leontius of Jerusalem or is just a contemporary. Source of quote: Leontius of Byzantium quoted on Leo Donald Davis, S. J., *The First Seven Ecumenical Councils (325–787): Their History and Theology* (Collegeville, MN: Liturgical Press, 1990), p. 232.

30 To Friends (January 26, 2007)

1. Pope Hormisdos (450–542 CE), Roman Catholic Pope in the period 514–523 CE.

31 To Lissa McCullough (March 13, 2007)

1. Thomas J. J. Altizer, "The Birth of the Vision," chapter 1 of *History as Apocalypse* (Albany, NY: State University of New York Press, 1985), pp. 17–30.
2. Mircea Elaide, (1907–1986), Romanian-born historian of religions, chair of History of Religions, University of Chicago, 1958–1986.
3. Wendy Doniger (b. 1940) has been Mircea Eliade Distinguished Service Professor of the History of Religions at the University of Chicago Divinity School since 1986.
4. See Thomas J. J. Altizer, *Living the Death of God* (Albany, NY: State University of New York Press, 2006) pp. 47–48.
5. D. G. Leahy, *Novitas Mundi: Perception of the History of Being* (1980; Albany, NY: State University of New York Press, 1994). See: http://dgleahy.com/p11.html

32 To Donald Weinstein (March 23, 2007)

*Donald Weinstein (b. 1926), professor in the Department of History, University of Arizona 1978–1992.

1. Perry Miller (1905–1963), Professor of History, Harvard University.
2. Hannah Arendt (1906–1975), German-Jewish political theorist and philosopher, noted academic and public intellectual in America after World War II. Here Altizer refers to Arendt's book, *On Revolution* (New York: Viking, 1963).
3. Thomas Jonathan "Stonewall" Jackson (1824—1863), Confederate general during the American Civil War and a staunch Presbyterian, of whom Altizer is a descendant and namesake.
4. *The Century of the Self* (BBC, 2002), a four-part documentary investigating the use of psychoanalysis by corporations and political parties to influence mass behavior.
5. H. L. Mencken, *The American Language* (New York: Alfred A.Knopf,1919).

33 To Friends (March 27, 2007)

1. Marcel Gauchet, *The Disenchantment of the World: A Political History of Religion* (Princeton: Princeton University Press, 1997); originally published in France as *Le désenchantement du monde: Une histoire politique de la religion* (Paris: Gallimard, 1985).
2. Gauchet (b. 1946), in the tradition of continental social scientists, combines being a philosopher, historian, and sociologist. He is Professor at the Centre de rescherches politiques Raymond Aron at Ecole des Hautes Etudes en Sciences Sociales (EHESS), Paris.

35 To John D. Caputo (May 17, 2007)

*John D. Caputo (b. 1940), Thomas J. Watson Professor emeritus, Religion and Humanities department, Syracuse University. An influential American postmodern religious thinker, he is described on the Syracuse University website as "a hybrid philosopher/theologian intent on producing impure thoughts, thoughts which circulate between philosophy and theology, short-circuits which deny fixed and rigorous boundaries between philosophy and theology" [http://asfaculty.syr.edu/pages/rel/caputo-john.html].

1. Thomas Sheehan (b. 1941), Professor of Religious Studies, Stanford University.
2. Julie Kristeva (b. 1941), Bulgarian-French intellectual and academic, notably in psychoanalysis, literary criticism, philosophy, and feminism.
3. Escaping from the house fire in 2004 that destroyed his library and many papers, Altizer jumped off the back porch, injuring his foot and back.

4. Caputo, "Atheism, A/theology, and the Postmodern Condition," in Michael Martin ed., *The Cambridge Companion to Atheism* (Cambridge: Cambridge University Press, 2007), pp. 276–282.
5. William J. Richardson, S. J. (1920), professor emeritus of Philosophy at Boston College. The paper to which this letter refers was shared privately (not for citation) by Richardson, "Whose Phenomenology?" a provisional draft of a work in progress presented at a symposium entitled "The Breakthrough of Phenomenology and Theology," Boston University, April 26, 2001.

36 To Brian Schroeder (June 7, 2007)

1. Brian Schroder, "Apocalypse, Eschatology, and the Death of God," in *Nietzsche and Levinas: After the Death of a Certain God*, eds. Jill Stauffer and Bettina Bergo (New York: Columbia University Press, 2009), pp. 232–248.
2. Levinas's term for what can be described as the remote otherness of God arising from a conception of the beyond expressed in terms of a third person, wherein the responsibility for the other takes place, a relation encountered as trace in the face of the other.

37 To John T. Wilcox (June 13, 2007)

*John T. Wilcox, professor emeritus of Philosophy, State University of New York at Binghamton.

1. *Trapeze* (MGM/UA, 1956), a circus film directed by Carol Reed, starring Burt Lancaster, Tony Curtis, and Gina Lollobrigida.
2. Friedrich Hölderlin (1770–1843), German poet who had a deep influence on Hegel, Schiller, Rilke, and Heidegger; he studied theology alongside Hegel and Schelling but turned to lyric poetry in undertaking similar questions of meaning as those expressed by Blake and Yeats. For Hölderlin, Christ is positioned as the last god of antiquity.
3. See Thomas J. J. Altizer, *Godhead and the Nothing* (Albany, NY: State University of New York, 2003) pp. 20–21.

38 To Friends (July 10, 2007)

1. J. N. Findlay, *Hegel: A Re-examination* (London: George, Allen and Unwin, 1958; New York: Macmillan, 1958). John Findlay (1903–1987) was a South African philosopher who taught in New Zealand, South Africa, England, and America, and gave the Gifford Lectures, 1964–1966.

Altizer wrote of Findlay: "I knew Findlay and admired him profoundly thinking that his Gifford Lectures are one of the great works of our era. And he is perhaps our most radical theological thinker and an ultimately radical Christian" (private correspondence, May 26, 2015).

39 To John D. Caputo (September 29, 2007)

1. See Chapter 11 of Altizer's *Living the Death of God* entitled "Prayer," pp. 143–153.

40 To Edward S. Casey (November 23, 2007)

*Edward S. Casey, Distinguished Professor of Philosophy at Stony Brook University. See: http://edwardscasey.com/

1. See Chapter 9 of Altizer's *Living the Death of God* entitled "Ethics," pp. 113–128.
2. Albert Schweitzer, *The Mysticism of Paul the Apostle*, trans. William Montgomery (1930; London: Black, 1931).
3. Thomas J. J. Altizer, *The Contemporary Jesus* (Albany, NY: State University of New York Press, 1997); see pp. 71–88.
4. Thomas J. J. Altizer, *History as Apocalypse* (Albany, NY: State University of New York, 1985).

41 To Friends (November 28, 2007)

1. Jeffrey Kluger, "What Makes Us Moral," TIME November 21, 2007. http://content.time.com/time/specials/2007/article/0,28804,1685055_1685076_1686619]
2. W. D. Hamilton (1936–2000), English evolutionary biologist, Royal Society Research Professor at Oxford University, who argued for a rigorous genetic basis for altruism.
3. Richard Dawkins (1941), English evolutionary biologist and "new atheism" author.
4. E. O. Wilson (1929), American biologist, environmentalist, and biodiversity researcher. University Research Professor Emeritus at Harvard.
5. Warder Allee (1885–1955), American zoologist and ecologist who promoted evolutionary group adaptation over individual evolutionary adaptation within the population of a species. This became known as the Allee effect or Allee principle.
6. George R. Price (1922–1975), American population geneticist and science journalist. Author of the Price equation, combining mathematics, biology and evolution, expressed a genetic basis for altruism. His suicide arose from his despair at attempting to undertake non-genetic altruism.

42 To Walter A. Strauss (December 4, 2007)

1. The letter to friends of November 28, 2007.
2. W. D. Hamilton, (2000) "My intended burial and why," *Ethology Ecology and Evolution* 12, no. 2 (2000): 111–122; p.122.

43 To Friends (October 30, 2008)

1. Richard Kearney, *The God Who May Be. A Hermeneutics of Religion* (Bloomington: Indiana University Press, 2001).
2. David Leahy (1937–2014), philosophical theologian, teacher of classics, philosophy, and religion at New York University and Loyola University, Maryland.
3. Karl Rahner, S. J. (1904–1984), Catholic theologian, prolific author and editor, who studied with Heidegger at Freiburg and worked with Ratzinger during Vatican II. For an accessible introduction to Rahner's demythologizing see Michael Barnes, "Demythologizing in the Theology of Karl Rahner," *Theological Studies* 55 (1994): 24–45.
4. Bernard Lonergan, S. J. (1904–1984), Canadian Catholic philosopher and theologian who developed his influential "general empirical method" to provide common ground to overcome the fragmentation of knowledge.
5. This refers to Bultmann's seminar series in Marburg in 1923.

45 To Friends (May 8, 2009)

1. Joachim Wach (1898–1955), German-born scholar of Jewish heritage but from a family long converted to Christianity, Wach emigrated to America in 1935 to teach at Brown University. He then taught at University of Chicago Divinity School, 1946–1955, and developed his sociology of religion as part a *Religionswissenschaft* understood as an empirical science separate from the philosophy of religion. Altizer studied under Wach in the History of Religions program at the University of Chicago.
2. Martin Bidley, *Blake and Goethe: Psychology, Ontology, Imagination* (University of Missouri Press, 1988).
3. Max Weber, *The Religion of China: Confucianism and Taoism* (1915).
4. Variously and inclusively meaning a cosmic mass, a glorification of existence, and the Infinite Passover of God; Altizer's use of the term is drawn from the work of D. G. Leahy.

46 To Friends (March 25, 2010)

1. Jacob Taubes (1923–1987), rabbi, philosopher, theologian, mystic, and eschatologist. Taubes, *Occidental Eschatology* (Stanford: Stanford University Press, 2009).

2. Thomas J. J. Altizer, ed., *Toward a New Christianity: Readings in the Death of God Theology* (New York: Harcourt, Brace, and World, 1967). The essay by Jacob Taubes "On the Nature of the Theological Method: Some Reflections on the Methodological Principles of Tillich's Theology" originally appeared in *The Journal of Religion* 34, no. 1 (January 1954): 12–25.
3. Jacob Taubes, *The Political Theology of Paul*, trans. Dana Hollander (Stanford: Stanford University Press, 2003). The lectures in this book were given in Berlin in February 1987, just before Taubes died of cancer.

47 To Friends (March 31, 2010)

1. Joachim of Fiore (c. 1130/35–1202 CE), apocalyptic thinker, theologian, and mystic from Calabria who, based on the Book of Revelation, conceived of a three-stage history of the word: first, that of the Father (the Old Covenant); second, that of the Son (the New Covenant of Christianity), and finally, that utopian age of the Holy Spirit that was expected to eventuate in 1260 CE.

50 To Friends (July 22, 2011)

1. *Ballets Russes,* directed by Daniel Geller and Dayna Goldfine (Zeitgeist Films, USA, 2005); see: http://www.balletsrussesmovie.com.
2. Established in Paris by Sergei Diaghilev in 1909, and later resident in Monte Carlo, the *Ballets Russes* performed in Europe and America (but never in Russia) until 1929 when, upon Diaghilev's death, it was disbanded.
3. *The Red Shoes* (1948), directed by Michael Powell and Emeric Pressburger.
4. Leonide Massine (1895–1975), a member of the *Ballets Russes* from 1914; Massine was a dancer and one of the leading choreographers of the twentieth century. See: http://massine-ballet.com/html/about_massine.ph
5. Jennifer Homans, *Apollo's Angels: A History of Ballet* (New York: Random House, 2010).
6. Moria Shearer (1926–2006), lead actress and dancer in *The Red Shoes.*
7. Sir Robert Helpmann (1909–1986), Australian ballet dancer, actor, director, and choreographer.

51 To Friends (July 23, 2011)

1. Max Weber, *The Protestant Ethic and the Spirit of Capitalism* (1905).
2. See Max Weber, *The Protestant Ethic and the Spirit of Capitalism,* trans. Talcott Parsons with a foreword by R. H. Tawney (New York: Charles Scribner's Sons, 1958), p.104.

52 To Andrew W. Hass (July 29, 2011)

1. See Thomas J. J. Altizer, *Living the Death of God* (Albany, NY: State University of New York, 2006) p. 9.
2. Werner Jaeger, *Aristotle: Fundamentals of the History of His Development*, trans. Richard Robinson (1923; New York: Oxford University Press, 1934).

53 To Friends (August 2, 2011)

1. Alfred Kazin (1915–1998), Jewish American writer and literary critic, appointed distinguished professor in the English department at Stony Brook in 1963. The letter refers to Edward Mendelson, "The Hidden Life of Alfred Kazin," *New York Review of Books*, August 18, 2011, a review of *Alfred Kazin's Journals*, selected and edited by Richard M. Cook (New Haven: Yale University Press, 2011).

54 To Friends (October 26, 2011)

1. Society for Phenomenology and Existential Philosophy, the 2011 annual meeting of which was held in Philadelphia, October 19–22.
2. Chapter 10 of Altizer's *Godhead and the Nothing* is entitled "The Body of the Abyss," pp. 143–158.

55 To Friends (November 14, 2011)

1. Occupy Wall Street began on September 17, 2011, as a protest against the Wall Street bailouts, global inequality, and the failure of democratic processes to deal with financial institutions' culpability for the 2008 economic collapse. See: http://occupywallst.org/

57 To Friends (February 11, 2012)

1. A paper to be presented at a conference entitled "The Crisis of American Democracy," organized by Professor Clark W. Butler, sponsored by the Indiana–Purdue Fort Wayne Human Rights Institute and Center for Applied Ethics, April 27–28, 2012, and circulated as a letter by Altizer.
2. Perry Miller (1905–1963), Professor of History, Harvard University
3. H. L. Mencken, *The American Language* (New York: Alfred A. Knopf, 1919).
4. William Faulkner (1897–1962), Nobel-prize winning novelist and short story writer from Oxford, Mississippi. His work, focused primarily on the imaginary Yoknapatawpha county, portrayed his vision of Southern life and society, especially its decay.

5. Henry James (1843–1916), American writer and later naturalized British citizen who wrote primarily of transatlantic culture focused on the dynamism, innocence, money, and spirit of the New World encountering, and often clashing with, the mores of the Old World.
6. William James (1842–1910), American philosopher, brother of Henry James, who wrote and taught on the intersections of philosophy, physiology, and psychology. His religious thinking in *The Varieties of Religious Experience* (1902) has been influential.
7. Eugene O'Neill (1988–1953), Nobel-prize winning American playwright.
8. Kirilov is the suicidal hero of Dostoevsky's novel *The Possessed,* also translated as *The Devils* or *Demons* (1872).

58 To Friends (March 2012)

1. William Hamilton (1924–2012), American radical theologian, who with Altizer co-authored *Radical Theology and The Death of God* (Indianapolis: Bobbs-Merrill, 1966). The response to this and the *Time* magazine cover story "Is God Dead?" (April 8, 1966, vol. 87, no. 14) led him to leave his tenured professorship at Colgate Rochester Divinity School. He taught at Portland State University until his retirement in 1986. See also Thomas J. J. Altizer, *Living the Death of God,* (Albany, NY: State Univeristy Of New York Press, 2006) pp. 12–13.
2. Hamilton wrote and hosted over forty episodes of *Look Up and Live* (CBS, in connection with the National Council of Churches). The Sunday morning series ran from 1954–1979.
3. See Ved Mehta, "The New Theologian," *New Yorker* magazine, Published as a three part series: November 13, 1965, pp. 63–153: November 20, 1965, pp.60–144; and November 27, 1965, pp.65–169. These were later published as Ved Mehta, *The New Theologian* (New York: Harper & Row, 1966).

59 To Friends (March 7, 2012)

1. Society of Biblical Literature and American Academy of Religion, respectively.
2. David E. Campbell and Robert D. Putnam, "God and Caesar in America: Why Mixing Religion and Politics Is Bad for Both," *Foreign Affairs* 91, no. 2 (2012): 34–43. See: https://www.foreignaffairs.com/articles/united-states/2012-02-12/god-and-caesar-america

60 To Ray L. Hart (May 13, 2012)

1. The term that philosophical theologian D. G. Leahy coined to describe an epochal new thinking that is reducible neither to theology nor to philosophy; see Leahy, *Foundation: Matter the Body Itself* (Albany, NY: State Univeristy of New York Press, 1996) pp. 5–6 and passim.

2. Auerbach, *Dante: Poet of the Secular World* [orig. 1929] (New York: New York Review Books, 2007).
3. That is, a uniquely Catholic "because," from the Latin, *quia*.
4. Pope Pius IX (1792–1878), pope in the period 1846–1878, his pontificate saw a notable shift to conservative Catholicism, including the dogma of the Immaculate Conception (1854) and the declaration of papal infallibility in the First Vatican Council (1869–1870).

61 To Friends (June 19, 2012)

1. Charles Stein, *Persephone Unveiled: Seeing the Goddess and Freeing Your Soul* (Berkeley: North Atlantic Books, 2006). For more on Stein see: http://www.charlessteinpoet.com.
2. Lucretius (Titus Lucretius Carus) (c. 99–c. 55 BCE), Roman Epicurean poet whose poem *De rerum natura* (On the Nature of the Universe), in its naturalistic discussion and description, is also the primary source of knowledge concerning Epicurean physics, and as such a major philosophical inspiration up to modern times.
3. Philip Larkin (1922–1985), English poet and librarian, jazz critic, and novelist.
4. W. B. Yeats (1865–1939), Irish poet and playwright, a Nobel-prize winner and major influence in twentieth-century poetry, member of the hermeticist group the Golden Dawn.
5. T.S. Eliot (1888–1965), American-English poet, playwright, and critic, a Nobel-prize winner and one of the major poets of the twentieth century, a convert to Anglo-Catholicism in 1927.
6. T. S. Eliot, *The Wasteland* (New York: Horace Liveright, 1922); originally published in the *New Criterion* (UK) and *The Dial* (USA). A response to what Eliot perceived as the crisis of modern life and culture, it became the most influential poem in defining Modernism.

62 To Friends (June 26, 2012)

1. Thomas J. J. Altizer, *Oriental Mysticism and Biblical Eschatology* (Philadelphia: Westminster, 1961).
2. Chapter 7 of Altizer's *The Apocalyptic Trinity* (New York: Palgrave Macmillan, 2012) is entitled "The Tragic Trinity," pp. 133–140.
3. Richard McKeon (1900–1985), influential American classical scholar, philosopher, and historian of ideas who taught for more than 40 years at the University of Chicago. See: http://www.richardmckeon.org/index.htm

63 To Friends (June 28, 2012)

1. See Thomas J. J. Altizer, *Living the Death of God*, pp. 9–10, and p. 11 for mention of Emory University's support during the death of God debate.

2. Rudolf Otto, *The Kingdom of God and the Son of Man: A Study in the History of Religion*, trans. Floyd F. Filson and Bertram Lee Woolf (1933; London: Lutterworth, 1938). Otto (1869–1937), German historian of religions, philosopher, and religious thinker is best known for *The Idea of the Holy* (1917), in which he coined the term *numinous* to describe the Holy.
3. Rudolf Otto, *Mysticism East and West: A Comparative Analysis of the Nature of Mysticism*, trans. Bertha L. Bracey and Richenda C. Payne (1926; London: Macmillan, 1932).
4. Shankara (788–820 CE), Indian mystic, philosopher, and saint whose writings are the foundation of the Advaita Vedanta school; he founded monastic orders and revitalized Hinduism against an increasingly dominant Buddhism.
5. Meister Eckhart (1260–1328), German Dominican priest, mystic, theologian, and philosopher.
6. Masao Abe (1915–2006), Japanese Zen Buddhist and religious studies scholar of the Kyoto school, which incorporated Western theology and philosophy into Buddhism, and a leader of Buddhist-Christian interfaith dialogue.
7. Keiji Nishitani (1900–1990), a leading Japanese Zen Buddhist scholar of the Kyoto School.
8. Herbert Richardson (b. 1932), contentious professor of religion and founder of Edward Mellen Press in 1972.

65 To Friends (November 9, 2012)

1. Oswald Spengler (1880–1936), German independent scholar and philosopher of history, best known for *The Decline of the West* (1918/1922), which viewed the First World War as a culmination of a historical-cultural process of civilizational cycles; he influenced both Heidegger and Wittgenstein.

66 To Friends (February 15, 2013)

1. Karl Barth, *Church Dogmatics*, vol. 2: *The Doctrine of God*, part 1, *The Knowledge of God: The Reality of God* [orig.1940] (Edinburgh: T&T Clark, 1957).

67 To Friends (June 7, 2013)

1. Stanford Encyclopedia of Philosophy, http://plato.stanford.edu, an online open-access encyclopedia created and maintained by Stanford University.
2. For a discussion of Altizer, the Eucharist, and religious language see J. Stephen Fountain, "Postmodernism, A/theology, and the Possibility of Language as Universal Eucharist," in *The Nature of Religious Language: A Colloquium*, Stanley E. Porter ed. (London: Bloomsbury, T & T Clark, 1996), pp. 131–47.

68 To Friends (August 6, 2014)

1. Krzstof Penderecki (b. 1933), acclaimed Polish composer and conductor whose work often focuses on an investigation of traumatic events. Penderecki's first opera, *The Devils of Loudon* (1968/69, 1972), with libretto by Penderecki via John Whiting's dramatization, is based on Aldous Huxley's 1952 nonfiction novel of religious and sexual hysteria, fanaticism, and supposed demonic possession in the French town of Loudon in 1634.
2. The Hamburg State Opera studio production (1969) see: http://www.naxos.com/catalogue/item.asp?item_code=101279
3. Penderecki, *St. Luke Passion* (1966) an avant-garde Passion noted for its atonality.
4. A 1932 opera based on the Book of Exodus by the Austrian Jewish expressionist and avant-garde composer Arnold Schoenberg (1874–1951).
5. There is a very brief mention; however there is a separate full Wikipedia article on the opera: http://en.wikipedia.org/wiki/The_Devils_of_Loudun_%28opera%29
6. An 1882 opera by Richard Wagner (1813–1883) concerning the quest of Parsifal for the Holy Grail and the holy spear.
7. Nietzsche's turn from championing Wagner to opposing him is set out in his essay *Nietzsche Contra Wagner* (1895). His turn from Wagner is twofold: Wagner's style of music is rejected as being against a Dionysian affirmation of life, and so makes Nietzsche feel sick, and Wagner's becoming an anti-Semitic Christian.

69 To Friends (September 22, 2014)

1. Despite being 80% Muslim, Indonesia recognizes Islam, Catholicism, Protestantism, Hinduism, and Buddhism as religions and is not an Islamic state; there has, however, been a recent rise of religious intolerance and persecution linked to Islamic sectarianism.
2. See for example the Pluralism Project at Harvard University, http://pluralism.org, and also, the Pew Research Center 2015 report on America's changing religious landscape, http://www.pewforum.org/2015/05/12/americas-changing-religious-landscape
3. Here Comes Everybody or HCE is the nom de plume of Humphrey Chimpden Earwicker, the main character of Joyce's *Finnegans Wake* (1939).
4. A play written by Eugene O'Neill in 1939, not performed until 1946, in which the central theme is the need for delusions to help us to cope with the painful reality of modern life, since without them we are reduced to nihilistic and suicidal despair.
5. James Baldwin, "Stranger in the Village" (originally published in *Harpers Magazine,* 1953), included in his *Notes of a Native Son* (Boston: Beacon, 1955). Baldwin (1924–1987), an American writer and social critic whose work critically investigates issues of race, sexuality, and class. He spent much of his life in Europe but his focus remained firmly on America, especially on experiences of inequality by black and gay Americans.
6. Heidegger's summer semester lecture course at University of Freiburg, 1935, not published in German until 1953 as *Einführung in die Metaphysik*. Available in

English as *Introduction to Metaphysics*, trans. Gregory Fried and Richard Polt (New Haven: Yale University Press, 2000).

7. Mississippi.
8. Major American writers from Mississippi include William Faulkner, Eudora Welty, Tennessee Williams, Richard Wright, Walker Percy, and Richard Ford.
9. A 12-bar blues song by Joe "King" Oliver, composed in 1928; it is the recording of June 28, 1928, by Louis Armstrong and his Hot Five that, with its solo introduction and up tempo, innovative technique, changed the way jazz was played. This 1928 recording is available at: https://www.youtube.com/watch?v=W232OsTAMo8.
10. Louis Armstrong (1901–1971), American jazz legend, trumpeter, singer, bandleader, and actor.
11. H. L. Mencken, *The American Language* (1919).
12. The Mexican-American War (1846–1848), conducted under the administration of President Polk was fought under the vision of Manifest Destiny, the American belief that they had a God-given right to expand their territory and "civilize" the continent. Also provoked by the decision of the recently Mexican territory of the independent Republic of Texas to join the Union in 1845, the war resulted in the United States gaining the territory that was to become California, Nevada, Arizona, New Mexico, and Utah.
13. It is estimated that while only about 1,500 American soldiers died as a result of battle, over 10,000 American soldiers died of illnesses including yellow fever, measles, mumps, and smallpox. Mexican casualties are usually estimated as at least 25,000.
14. Charles Baudelaire (1821–1867), French poet, art critic, essayist, and translator of Edgar Allan Poe, who had a major impact on the rise of modernism, especially his "The Painter of Modern Life" (1859), in which modernity is defined as a combination of the eternal and the fleeting, and his transgressive collection of poems, *Les Fleurs du mal* (*The Flowers of Evil*) in 1857.
15. *The Air Conditioned Nightmare* (1945) recounts the three-year journey made by the American writer Henry Miller (1891–1980) on his return to America in 1939 after ten years abroad. In the book Miller stringently critiques what he experiences as the soullessness, materialism, and spiritual-cultural vacuum of America.

71 To Friends (March 2, 2015)

1. William Blake engraved 22 *Illustrations to the Book of Job* (1826).For the complete collection, see: http://www.blakearchive.org/exist/blake/archive/copy.xq?copyid=bb421.1&java=noFor the plate referred to, see: http://www.blakearchive.org/exist/blake/archive/object.xq?objectid=bb421.1.spb.13&java=no

73 To Friends (April 11, 2015)

1. Jacob Taubes, "On the Nature of the Theological Method: Some Reflections on the Methodological Principles of Tillich's Theology," cited in the letter of March 25, 2010, on Taubes.

2. See Thomas J. J. Altizer, *Living the Death of God* (Albany, NY: State University of New York Press, 2006), p. 10.
3. Held at University of Chicago in the autumn of 1964, the History of Religions and Systematic Theology seminar included Tillich's last public lecture on October 11, 1965, "The Significance of the History of Religions for the Systematic Theologian," after which he suffered what was to be a fatal heart attack, dying eleven days later. The conference proceedings are published in *The Future of Religions,* ed. Jerald C. Brauer (New York: Harper & Row, 1966); Tillich's final lecture appears on pp. 80–94.
4. See John Luther Adams, *Paul Tillich's Philosophy of Culture, Science, and Religion* (New York: Harper & Row, 1965), and *The Thought of Paul Tillich*, ed. James Luther Adams, Wilhelm Pauck, and Roger Lincoln Shinn (New York: Harper Collins, 1985).

74 To Friends (April 26, 2015)

1. Cyril O'Regan, "Benedict the Augustinian," in *Explorations in the Theology of Benedict XVI,* John C. Cavadini ed., (South Bend: University of Notre Dame Press, 2013), pp. 21–62.
2. Altizer wrote his master's thesis on "Nature and Grace in the Theology of Saint Augustine" (University of Chicago, 1951).

Recipients

John D. Caputo, Thomas J. Watson Professor Emeritus, Religion and Humanities, Syracuse University, USA.

Edward S. Casey, Distinguished Professor of Philosophy at State University of New York at Stony Brook, USA.

John B. Cobb, Jr., American process theologian, philosopher, and environmental ethicist; cofounder of the Center for Process Studies in Claremont, California, USA.

Alina N. Feld, author of *Melancholy and the Otherness of God: A Study of the Hermeneutics of Depression* (2011), teaches at Hofstra University and General Theological Seminary, USA.

Gan Yang, former editor of the influential book series *Culture: China and the World* published in Beijing in the 1980s; he studied at the University of Chicago, USA, in the 1990s, and has since been based in Hong Kong as a research scholar and academic administrator.

Ray L. Hart, Professor of Religion and Theology, Boston University School of Theology, USA.

Andrew W. Hass, Reader in Religion at the University of Stirling, Scotland, UK.

D. G. Leahy (1937–2014), philosophical theologian, professor of classics and religious studies at New York University, USA, distinguished visiting professor at Loyola University of Maryland, USA, and founder of New York Philosophy Corporation.

Lissa McCullough, independent scholar, co-editor with Brian Schroeder of *Thinking Through the Death of God: A Critical Companion to Altizer* (2004), and author of *The Religious Philosophy of Simone Weil* (2014).

Cyril O'Regan, Huisking Professor of Theology, University of Notre Dame, USA.

Brian Schroeder, professor in the Department of Philosophy at Rochester Institute of Technology, USA.

Walter A. Strauss, late scholar of comparative literature, Treuhaft Professor of the Humanities Emeritus, Case Western Reserve University, USA.

Donald Weinstein (1926–2015), specialist in the Italian renaissance, formerly professor in the Department of History, University of Arizona, USA, until his retirement in 1992.

John T. Wilcox, late professor of philosophy, State University of New York at Binghamton, USA.

Edith Wyschogrod (1930–2009), American philosopher of religion, J. Newton Rayzor Professor of Philosophy and Religious Thought at Rice University, USA.

Index